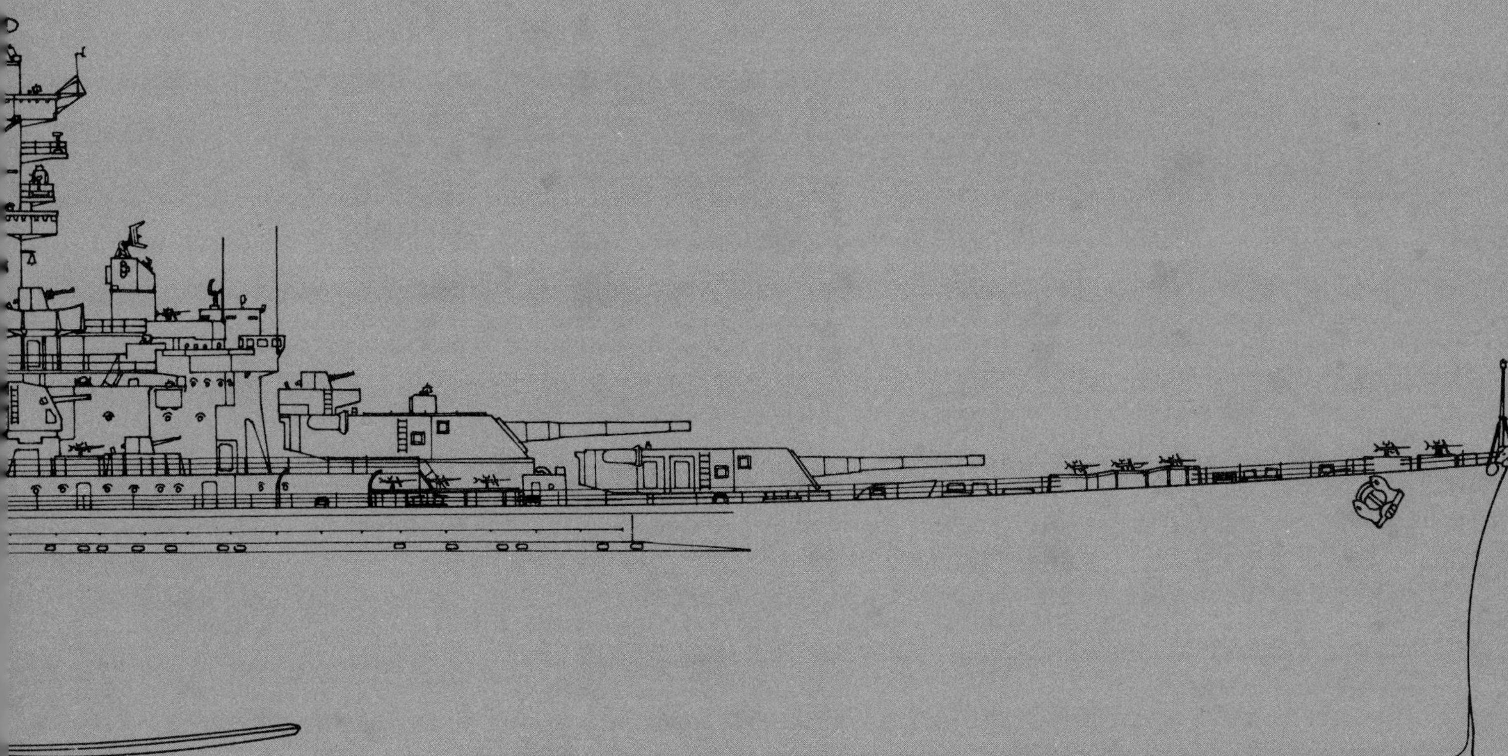

The Illustrated Encyclopedia of 20th Century WEAPONS AND WARFARE

692

VOLUME 1
A1/AMX

The Illustrated Encyclopedia of 20th Century

WEAPONS AND WARFARE

COLUMBIA HOUSE/New York

Editor: Bernard Fitzsimons
Designer: David Harper
Editorial Assistants: Suzanne Pearce
Robin Cross
Sarie Forster
Will Fowler
Production: Tony Holdsworth
Picture Research: Diane Rich
Editorial Director: Graham Donaldson

BERNARD FITZSIMONS, General Editor
ANTONY PRESTON, Naval Consultant
BILL GUNSTON, Aviation Consultant
IAN V. HOGG, Land Weapons Consultant

JOHN BATCHELOR, Illustrator

Cover Design: Harry W. Fass
Production Manager: Stephen Charkow

Distributed by Columbia House, 51 West 52nd Street,
New York, New York 10019
Printed in the United States of America

"My argument is that War makes rattling good history. . . ."

Thomas Hardy, *Spirit Sinister*

INTRODUCTION

Any profound understanding of the wars which have shaped our world depends on an understanding of the weapons with which they were fought. In these volumes, details of all the weapons of our century are gathered together in a single work.

The facts and information packed into these magnificent volumes could only be duplicated in a library of reference works, and even then, so thorough has been the work of the contributors that new facts are constantly being brought to light, new theories advanced and old myths exploded.

COVERAGE

Almost any weapon used in substantial numbers by regular armed forces during this century will be found in these pages. In addition, weapons which have had an impact on the development of modern armaments without actually seeing service, and even some projects which were never realized but are notable for their influence on subsequent developments, will be included. Entries will be illustrated wherever possible, with photographs, diagrams and color illustrations combining to provide a complete picture of modern weaponry in all its forms.

WARSHIPS

Ships appear under name, capital ships having their own entries where possible and smaller warships being discussed in classes under the name ship of the class. The *Bismarck,* for example, will have its own entry, but the cruisers *Juneau, San Diego* and *San Juan* are accounted for under the *Atlanta class.*

AIRCRAFT

Military aircraft of all types are included, tactical transports and trainers as well as bombers and fighters. For the most part they appear under their official name—P-51 will be cross-referenced to *Mustang*—or designation where no official name was assigned. Soviet aircraft are entered under their NATO code names (those in service since 1954); numbered types appear under the manufacturers. Again, cross-indexing by manufacturer and service designation will ensure that any plane can be located with maximum ease.

TANKS AND APCS

The main mobile combat weapons will appear under their official name designations—PZKPFW 1, *T-34, Churchill* and so on.

SMALL ARMS

Small arms of all types appear in the A-Z listed under manufacturer's (*Browning, Mauser*) or designer's (*Nambu, Degtyarev*) name, or under the official service name (*Bren, AAT-52*) where this is better known.

MISSILES

Missiles of all types are covered, from giant multi-stage ICBMs to the smallest man-portable antitank rocket launcher. Again, they appear under their official names, or designations where they are not named, with Soviet missiles under their NATO code names.

ARTILLERY

A special artillery section will cover all the main ordnance used by the major powers during this century. Starting with German light flak, taper-bore antitank, mountain and infantry guns, it will continue through medium and heavy flak, PAK and field artillery, recoilless, railway, coastal guns, ammunition and naval ordnance, providing a complete survey of 20th century artillery. In addition, individual weapons of particular note—the German "Bruno" railway guns, for example, will have their own, more detailed entries in the main A-Z section of the work.

CONTRIBUTORS

Authority and readability are the keynotes of the encyclopedia. All entries and specifications are checked by a panel of consultants who have been chosen for their long involvement in, practical experience of, and, above all, enthusiasm for, their particular subject.

The contributors, too, are enthusiasts to a man. Many have devoted a large part of their lives to the study of their field, and their combined knowledge and experience has added immensely to the information we present here.

A

A Class British destroyers See ***Acasta***

A Class Danish submarines See ***Havmanden***

A 1

First British A Class submarine. The A Class, of 13 boats, was designed by Vickers as an improved version of the US *Adder* Class, and was the first submarine class designed in Britain. *A 1* (ordered as Holland No 6) was launched in July 1902 and *A 13*, the first British submarine with a diesel engine, was launched in April 1905.

On March 18, 1904 *A 1* was sunk off Portsmouth in a collision with the liner *Berwick Castle*. She was later raised and put back into service but she was the first to be disposed of, as she was sunk as a target in 1911, followed by *A 3* in 1912. In January 1914 *A 7* was lost in Whitesand Bay when she dived into the mud.

During the First World War the survivors were used for harbour defence, *A 2, A 4, A 5* and *A 6* at Portsmouth, *A 8* and *A 9* at Devonport and *A 10, A 11* and *A 12* at Ardrossan. *A 13* had an experimental heavy oil (diesel) engine which was unreliable and so she was laid up in October 1914. The others were reduced to training by 1918, and were sold in 1919-20, except *A 2* which was wrecked while awaiting disposal and was not sold until 1925.

Displacement: (A 1) 185/203 tons surfaced/submerged; *(A 2-13)* 190/205 tons surfaced/submerged *Dimensions: (A 1)* 31.47 m (103 ft 3 in) oa×3.58 m (11 ft 9 in)×3.05 m (10 ft); *(A 2-13)* 32 m (105 ft) oa×3.85 m (12 ft 7¾ in)×3.22 m (10 ft 7 in) *Machinery: (A 1* surfaced) 1-shaft Wolseley petrol engine, 400 bhp=10.4 knots; *(A 1* submerged) electric motor, 150 ehp=6 knots; *(A 2-4* surfaced) 1-shaft Wolseley petrol engine, 450 bhp=11 knots; *(A 2-4* submerged) electric motor, 150 ehp=6 knots; *(A 5-12* surfaced) 1-shaft Wolseley petrol engine, 600 bhp=11.4 knots; *(A 5-12* submerged) electric motor, 150 ehp=6 knots; *(A 13* surfaced) 1-shaft Vickers diesel engine, 500 bhp=11.4 knots; *(A 13* submerged) electric motor, 150 ehp=6 knots *Armament: (A1)* 1×18-in (460-mm) torpedo tube (3 torpedoes); *(A2-13)* 2×18-in torpedo tubes (4 torpedoes) *Crew:* 11

A 1

German torpedo boat class. When the German army occupied the Flanders coast in the early weeks of the First World War the navy moved a number of light units to the new bases to attack British and French shipping. The Marineamt designed a new class of single-funnelled torpedo boats or light destroyers specially for operations off the Flanders coast, with its shallow waters and sandbanks, and these were given 'A' numbers to avoid confusion with older boats.

The boats varied in size, from the 109-ton *A 1-25* group to the 330-ton *A 96-113* group ordered in 1918, but they were all similar, with a single funnel, raised forecastle, and a pair of 18-in torpedo tubes amidships. The final groups had a stronger gun armament and were fitted for minesweeping, and from *A 26* onwards oil fuel replaced the earlier coal.

Most of the boats stricken in 1920-22 were surrendered to Britain and scrapped there or in designated Allied ports. *A 4-16, A 20, A 30* and *A 40-47* were handed over to Belgium as reparations after the First World War and received numbers A1PC-A25PC; *A 17* and *A 21* were sunk during the Kapp Putsch in 1920; *A 32* ran aground in the Baltic in 1917 and was salvaged by Estonia as the *Sulev*. *A 59, A 64, A 68* and *A 80* were taken over by Poland and renamed *Slazak, Krakowiak, Rujawiak* and *Goral*.

A 1-25 Group:
A 2 and *A 6* were sunk by British destroyers May 1, 1915; *A 15* sunk by British destroyers August 23, 1915; *A 7* and *A 19* sunk by British and French destroyers March 21, 1918; *A 3* lost 1915; *A 10* mined 1918; *A 13* bombed in dock 1917; *A 1, A 17, A 18, A 21-25* stricken 1921-22; *A 4-5, A 8-9, A 14, A 16* stricken 1927. *A 40* survived the Second World War and was scrapped in 1948.

Displacement: 109 tons normal *Length:* 41.58 m (136 ft 5 in) oa *Beam:* 4.6 m (15 ft 1 in) *Draught:* 1.52 m (4 ft 11¾ in) max *Machinery:* Single-shaft triple-expansion, 1200 hp=20 knots *Armament:* 1 50-mm or 52-mm (2-in); 2 45-cm (17.7-in) torpedo tubes; 4 mines *Crew:* 28

A 26-55 Group:
A 26-29, A 31, A 33-39, A 41, A 44-46, A 48-49, A 52-55 stricken 1920-21; *A 30, A 40, A 42, A 47* stricken 1927; *A 32* given to Russia in 1940; *A 43* scrapped 1943; *A 50* mined 1917 and *A 51* scuttled 1918.

Displacement: 227-229 tons normal *Length:* 50 m (164 ft 0½ in) oa *Beam:* 5.32-5.62 m (17 ft 5½ in-18 ft 5¼ in) *Draught:* 2.34 m (7 ft 8 in) max *Machinery:* Single-shaft geared turbine, 3250 hp =25 knots *Armament:* 2 88-mm (3.5-in); 1 45-cm (17.7-in) torpedo tube; fitted for minelaying *Crew:* 29

A 56-67 Group:
A56, 57, 58 mined 1918; *A59* scrapped 1927; *A60* mined 1917; *A61, 62* ceded to Britain 1920 and scrapped 1923; *A63, 66* ceded to France 1920 and scrapped 1923; *A64* ceded to Poland 1920 and scuttled off Danzig 1939; *A65* ceded to Brazil 1920 and broken up in Britain; *A67* scrapped incomplete 1921.

A 68-79 Group:
A 69-70, A 74-76 and *A 78* stricken 1920; *A 71-73, A 77* and *A 79* mined 1918; *A 68* scuttled 1939 off Danzig.

(A56-79) *Displacement:* 330-335 tons normal *Length:* 60-61.1 m (196 ft 10 in-200 ft 5½ in) oa *Beam:* 6.42 m (21 ft 0 in) *Draught:* 2.34 m (7 ft 8 in) max *Machinery:* 2-shaft geared turbines, 6000 hp=28 knots *Armament:* 2 88-mm (3.5-in); 1 45-cm (17.7-in) torpedo tube *Crew:* 50

A 80-91 Group:
A 81, A 86-91 stricken 1920; *A 82* scuttled Fiume 1918; *A 80* scrapped 1938; *A 83-85* scrapped incomplete 1919.

Displacement: 330 tons normal *Length:* 60.37 m (198 ft 0¾ in) oa *Beam:* 6.41 m (21 ft 0¼ in) *Draught:* 2.11 m (6 ft 11 in) max *Machinery:* 2-shaft geared turbines, 5700 hp=26 knots *Armament and crew:* As *A 68-79*

A 92-113 Group:
A 92-95 stricken 1920; *A 96-113* scrapped while on stocks 1919.

***A 12* of the British A Class of submarines running on the surface with her crew on deck**

PPL

Displacement: 335 tons normal *Length:* 61.2 m (200 ft 9½ in) oa *Beam:* 6.42 m (21 ft 0¾ in) *Draught:* 2.12 m (6 ft 11½ in) max *Machinery:* 2-shaft geared turbines, 6000 hp=26½ knots *Armament and crew:* As *A 68-79*

A 1

Italian midget submarine class. In 1912-13 the Italian navy built two experimental midgets for harbour defence at Venice. They were known unofficially as *Alfa* and *Beta* as they were never part of the navy, and were scrapped in 1915-16.

Profiting by this experience Lieutenant-General Ferrati designed a class of small electric-driven submersibles for defending the Adriatic harbours. The plans of the French *Naïade* Class were used as a basis, and like them they carried two 45-cm (17.7-in) torpedoes in external drop-collars. All five were built by the Arsenal at La Spezia, being laid down in July 1915 and completed between December 1915 and March 1916. The boats were numbered *A 1-6* and were discarded in September 1918. Submerged endurance was about 13.5 km (8½ miles) at a speed of 4½ knots.

Displacement: 31.25/36.7 tons surfaced/submerged *Length:* 13.5 m (42 ft 10 in) *Beam:* 2.218 m (7 ft 2 in) *Draught:* 2.275 m (7 ft 3 in) *Machinery:* Single-shaft electric motor, 46-60 ehp=6.8 knots (surfaced), 5.08 knots (submerged) *Armament:* Two 45-cm (17.7-in) torpedoes *Crew:* 4

A 1

First Norwegian A Class submarine. The four boats in this class were built for Norway by Krupp's Germania yard in 1907-14. The prototype was named *Kobben* but the class was given numbers on completion. They were very similar to the original German U-Boat *U 1* and the Austro-Hungarian *Ub 1* and *Ub 2* which immediately preceded them.

PPL

Kobben (A 1), first of the Norwegian A Class submarines. Note the external torpedo tube

Many improvements were incorporated into *A 2-4* which were the first submarines built in Germany with diesel engines. *A 1* had the Körting kerosene motor which was to continue in service in the German navy for some years, and she also had an external torpedo-tube at the stern, whereas her sisters had an internal tube. The later boats were fitted with a manoeuvring propeller in the bow (a most unusual feature in submarines) and also with radio equipment.

A fifth unit, *A 5* was still under construction at Kiel when war broke out in August 1914, and so she was incorporated into the German navy as *UA*. From 1916 she served only on training duties, and after the Armistice she was scrapped in Britain. The other four had been laid up in 1940.

Displacement: (*A 1*) 206/259 tons surfaced/submerged; (*A 2-4*) 268/355 tons surfaced/submerged *Length:* (*A 1*) 34.25 m (112 ft 4½ in); (*A 2-4*) 46.47 m (152 ft 5½ in) *Beam:* (*A 1*) 3.7 m (12 ft 1½ in); (*A 2-4*) 4.78 m (15 ft 4¼ in) *Draught:* (*A 1*) 2.95 m (9 ft 8 in); (*A 2-4*) 2.7 m (8 ft 10 in) *Machinery:* (*A 1* surfaced) 2-shaft Körting oil engine, 450 bhp=11.9 knots; (*A 1* submerged) electric motors, 300 ehp=8.9 knots; (*A 2-4* surfaced) 2-shaft Krupp diesel engines, 700 bhp=14 knots; (*A 2-4* submerged) electric motors, 380 ehp=9 knots *Armament:* 3×18-in (46-cm) torpedo tubes (2 bow, 1 stern) 4 torpedoes (five in *A 2-4*) *Crew:* 12 (*A 2-4* had 16)

A-1 US Navy attack bomber See **Skyraider**

A1-A7 US submarines See ***Adder***

The A1N2, second Japanese version of the Gloster Gambet, had the original Nakajima-built Jupiter VI replaced by the 520-hp Nakajima Kotobuki 2, a nine-cylinder air-cooled radial driving a two-bladed metal propeller. Fifty A1N1s and 100 A1N2s were built

In 1916 the Norwegian navy bought Farman floatplanes for trials, and when they broke down or ran out of fuel, submarines were able to recover them by surfacing gently underneath. Here a recovered seaplane is hoisted off the deck of the *A 4*. Note the spare floats lashed to the deck casing

Norwegian Navy

A1N/A2N/A4N Nakajima

Japanese navy biplane fighters. To build up their aeronautical expertise after the First World War, the Japanese collaborated with the more experienced procurement and design staffs of foreign countries. The most important of these countries was Britain, and the leading fighters of the Imperial Japanese Navy in the 1920s stemmed from the British Gloucestershire (later Gloster) Aircraft company in collaboration with Nakajima. The first model was the Gloucester-built Mars II, a single-seat biplane with 230-hp Bentley rotary, of which 30 were delivered to Japan in 1922 as the Sparrowhawk. This was followed by the Gambet, built as the Nakajima A1N1 and powered by the Nakajima-built Bristol Jupiter of 436 (later 450) hp. Similar to the RAF's Grebe, the A1N had two Vickers guns, could carry four 9-kg (20-lb) bombs and in a second model had a sliding canopy.

The same engine powered the A2N, another single-seater, of which 100 were built between 1930 and 1933, but this was based on US designs (Boeing and Curtiss) and was visibly noteworthy for its pointed elliptical wings. Extremely manoeuvrable, these popular machines served aboard the carriers *Kaga* and *Akagi.*

No fewer than 300 of the final biplane naval fighter, the A4N, were built in 1935-37. Developed from the A2N, this was the first Japanese navy fighter of indigenous design. Powered by a 770-hp Nakajima Hikari, A4Ns served in the Chinese war but were rated inferior to the A2N except for speed.

Span: (*A1N*) 9.70 m (31 ft 10 in); (*A2N*) 9.39 m (30 ft 10 in); (*A4N*) 9.96 m (32 ft 8 in) *Length:* (*A1N*) 50 m (21 ft 4 in); (*A2N*) 6.58 m (21 ft 7 in); (*A4N*) approx. 7 m (23 ft) *Gross weight:* (*A1N*) 1000 kg (2205 lb); (*A2N*) 1300 kg (2866 lb); (*A4N*) approx. 1360 kg (3000 lb) *Max speed:* (*A1N*) 219 km/h (136 mph); (*A2N*) 323 km/h (201 mph); (*A4N*) 352 km/h (219 mph)

Note: Japanese aircraft designations were complex. Prototypes were given a 'Ki' number (thus, Ki-21) by the army and a number based on the Japanese year 2585 (our 1925) by the navy (thus, Type 15=2600=1940). Production army types retained the Ki number, with added suffix numbers for subsequent models, plus a Type number based on the year (thus, Type 99=2599=1939). Navy designation resembled the old US Navy system with a letter for the duty (eg, A, carrier fighter; B, carrier attack; C, carrier reconnaissance), followed by a number and letter for the builder. Thus the above entry describes the first, second and fourth carrier fighters from Nakajima.

A2F US Navy attack bomber See ***Intruder***

A-3, A3D US Navy attack bomber See **Skywarrior**

A3J US Navy attack bomber See **Vigilante**

A-4

German ballistic missile. This large liquid-fuel rocket was developed at the experimental rocket station at Peenemünde on the Baltic coast of Germany between 1938 and 1944. Originally designed for use by the German army to attack battlefield rear areas beyond the range of conventional field artillery, its size was limited by a requirement for the weapon to pass through railway tunnels on flat-wagons.

From nose to tail the A-4 comprised: a one-tonne amatol high-explosive warhead, guidance compartment with three-axis gyro-stabilizer, ethyl alcohol-water fuel tank, liquid oxygen tank, turbopumps, hydrogen peroxide tank for the steam generator (which drove the turbopumps), and a combustion chamber of 25 000 to 26 000-kg (55 110 to 57 320-lb) thrust. Control was applied by gyroscopes which worked four graphite vanes in the exhaust and aerodynamic rudders on the fins.

When test firings began at Peenemünde in 1942 the first A-4 ignited but failed to develop sufficient thrust due to a fuel feed problem; instead of taking off it toppled onto its side and exploded. The second A-4 broke up in flight. However, the third rocket launched on October 3, 1942 was a complete success impacting on the Baltic some 190 km (118 miles) east of the launch site. The apogee of the trajectory was about 85 km (53 miles).

Before proper trials could be completed, Hitler ordered the rocket into large scale production—but now, as a 'revenge weapon' against London and the Home Counties. It then became known as the V-2 from the German *Vergeltungswaffe*, or revenge weapon. A large production plant was built south of Peenemünde and three more factories were set up—near Vienna, in a suburb of Berlin, and in the former Zeppelin hangars at Friedrichshafen.

In the meantime the Royal Air Force had obtained reconnaissance photos of Peenemünde which showed unmistakable signs of rocket activity. Six hundred bombers made a night attack on August 17, 1943, but although badly mauled much of the station, including the complicated test stands, the guidance and control laboratory and the wind tunnel, was left undamaged.

Mindful of growing Allied air power the German High Command ordered a disused oil depot south of the Harz mountains to be converted into a gigantic underground factory—the so-called *Mittlewerke*—from which, six months later, A-4 rockets began to flow at the rate of 300 a month.

Operationally, rockets were transported on *Meillerwagen* trailer-erectors and deployed by specially-trained army units. It was possible to launch from any reasonably flat ground and rockets were often set up among trees to avoid detection from the air.

The attack on London began on September 8, 1944 from sites at Wassenaar near The Hague in Holland. The first rocket fell in Croydon, killing three civilians and injuring 17 others. Since it approached at supersonic speed, there was no warning, and while the relatively weak explosive used meant its effects were limited in scope, the lack of warning added to the confusion caused by the random explosions.

The erection procedure was relatively simple. The rocket was elevated by hydraulic jacks on the transporter-erector and set upright on its fins on a simple launch table. It was then aligned by the launch crew using a ratchet device which turned the rocket bodily on the launch table until the elevator fins were in a plane normal to the target. After the rocket had been fuelled from road tankers it was fired from a safe distance by the battery commander in an armoured vehicle which

IWM

The A-4 rocket was thoroughly tested by the Allies after the Second World War: here a captured and repainted example is shown on the *Meillerwagen* transporter and during pre-launch erection, which was carried out by hydraulic jacks attached to the vehicle. The use of mobile transporter/launchers gave A-4s much greater flexibility than V-1s, which depended on concrete launching ramps

IWM

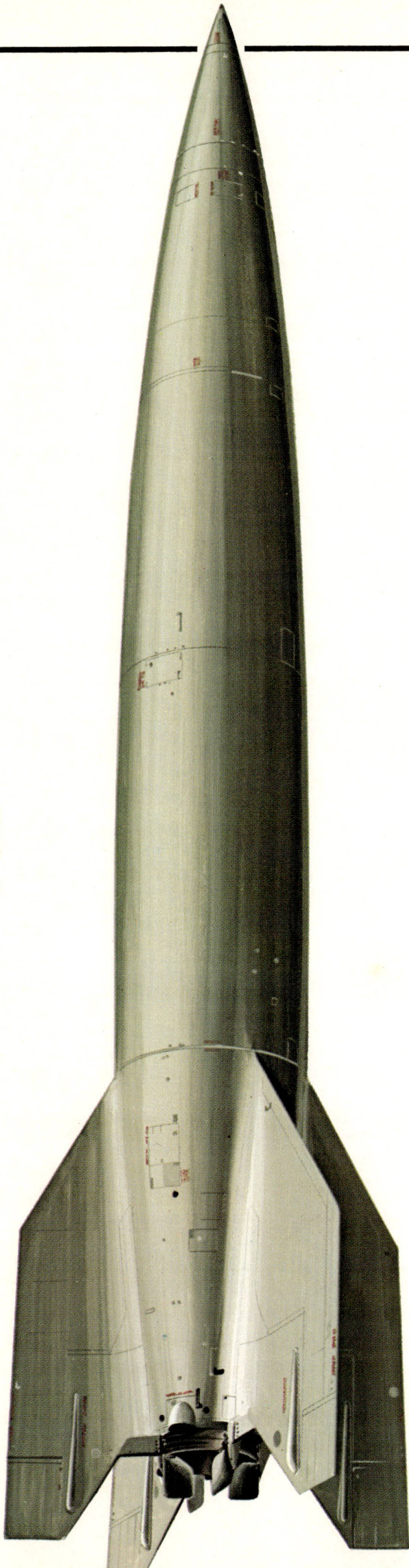

Popperfoto

Werner von Braun (above) was one of the leading scientists in the German rocket programme which included the A-4 (left). *Below:* A-4 being launched by Allied scientists during postwar tests with the missile

Imperial War Museum

I V Hogg

was linked to the rocket by an umbilical cable.

A total of 4320 A-4s were fired between September 6, 1944 and March 27, 1945, including 1120 against southeast England and 2500 against continental targets (including the town of Antwerp), the remainder being used for test and training purposes. Civil Defence records show that some 1050 actually fell on England. However, many rockets went astray and fell into the sea or exploded in flight. Some failed at, or shortly after, takeoff causing casualties among the launch crews.

Two A-4b rockets (improved A-4s fitted with swept-back wings and enlarged aerodynamic rudders) were launched on test in Poland in the winter of 1944-45 with the aim of stretching the range to 750 km (466 miles). Although one rocket flew with some success the project was discontinued in order to focus maximum attention on the A-4.

Length: 14 m (46 ft) *Diameter:* 168 cm (66 in) *Fin span:* 3.57 m (11.7 ft) *Launch weight:* 12 870 kg (28 380 lb) *Range:* 306-320 km (190-200 miles)

A-4, A4D US Navy attack bomber See **Skyhawk**

A-5 US Navy attack bomber See **Vigilante**

A5M Mitsubishi

Japanese navy fighter. Designed and developed entirely by Jiro Horikoshi's team at the vast Mitsubishi Heavy Industries, which soon was to create the famed Zero, the A5M was one of the most important warplanes of the 1930s. It was an outstanding example of the philosophy followed by Japanese designers, and passionately believed by that country's pilots, that the things that mattered in air combat were manoeuvrability and pilot view. As a result the A5M, powered by a 550-hp Nakajima Kotobuki (Jupiter), was a trim but very light machine with an elliptical wing (it was the Imperial Navy's first monoplane), open cockpit, spatted landing gear and twin Vickers 0.303-in guns.

Most production models, called Type 96 and code-named 'Claude' by the Allies in the Second World War, had a 610-hp Kotobuki 3 and could carry two 30-kg (66-lb) bombs. No fewer than 982 (possibly a few more) were built by Mitsubishi and the Sasebo navy yard in 1935-39. They were used in violent air battles against both the Chinese and the Soviet Union between 1937 and 1940, acquitting themselves extremely well even against the I-15 and I-16. Hundreds served in front-line squadrons from carriers and shore bases, and they were still the main navy fighter type at the time of Pearl Harbor.

Span: 11 m (36 ft 1 in) *Length:* 7.67 m (24 ft 9½ in) *Gross weight:* 1707 kg (3763 lb) *Max speed:* 439 km/h (273 mph)

Note: Code-names were assigned by the Allies in 1942-45 to overcome the difficulty of discovering or remembering correct Japanese designations. In general, fighters received boys' names and other aircraft girls', the names being chosen for clear intelligibility.

A-6 US Navy attack bomber See **Intruder**

This diagram shows the main components of the A-4's propulsion and guidance systems

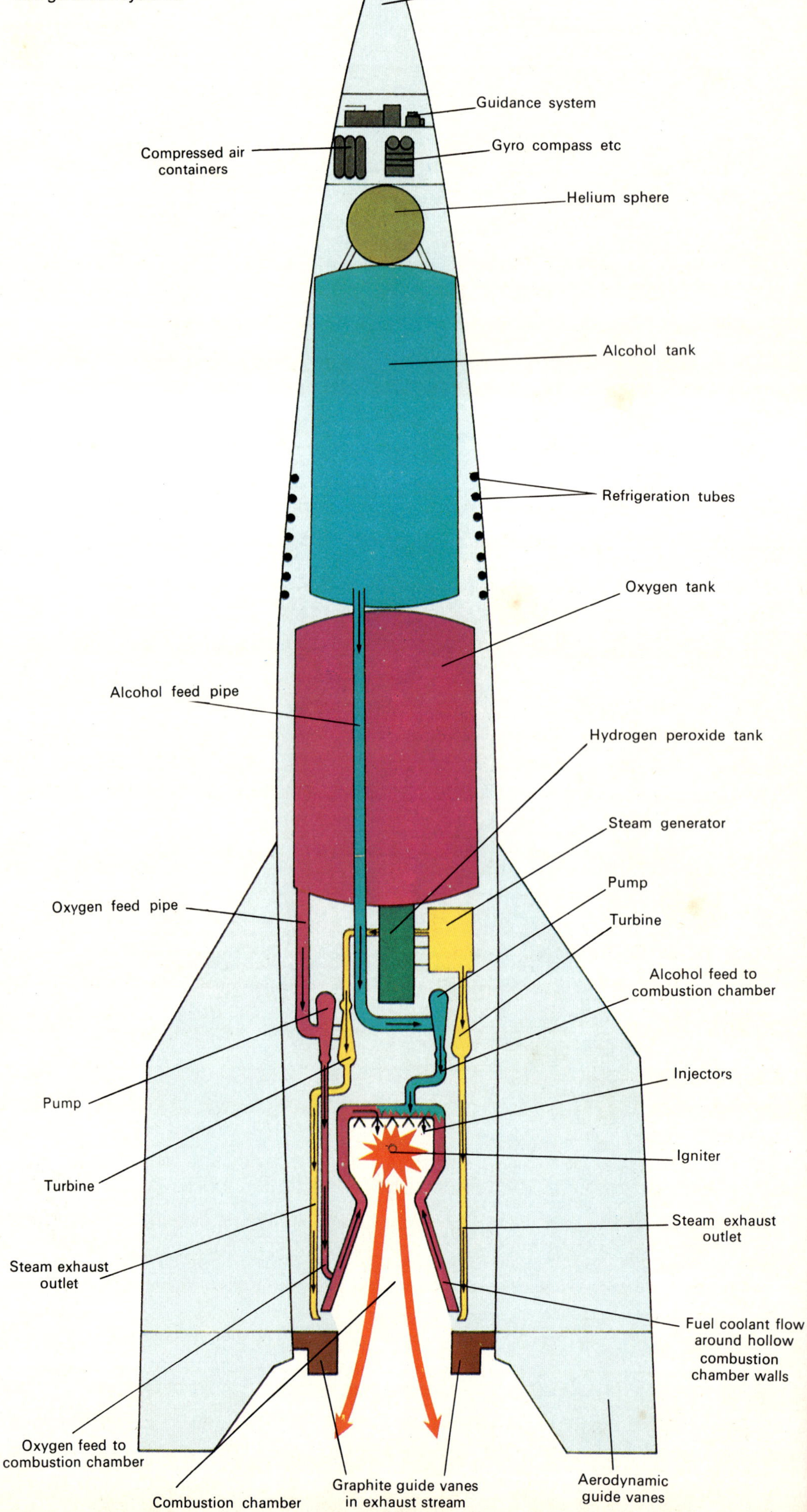

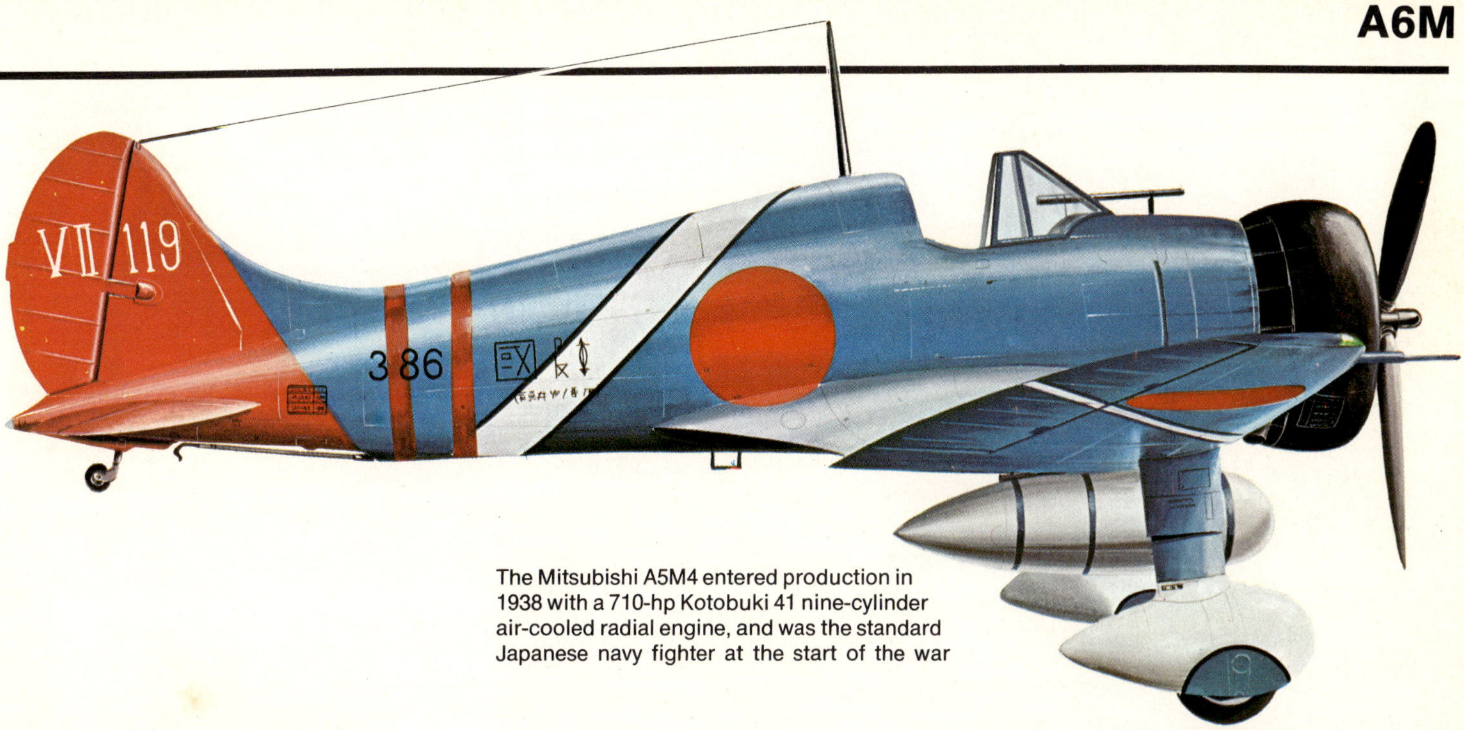

The Mitsubishi A5M4 entered production in 1938 with a 710-hp Kotobuki 41 nine-cylinder air-cooled radial engine, and was the standard Japanese navy fighter at the start of the war

A6M Zero-sen Mitsubishi

Japanese navy fighter. By far the most important Japanese warplane of the Second World War, the Mitsubishi A6M—code-named 'Zeke' by the Allies, but universally known as the Zero—began the war by demolishing all opposition, and it succeeded in creating a myth of invincibility that has seldom been achieved by any weapon. It was designed by Horikoshi's team at Nagoya, the rival Nakajima company having decided that the navy's October 1937 specification could not possibly be met.

The first A6M1 prototype was flown at the navy's Kasumigaura airfield on April 1, 1939, and proved a winner from the start. Though it had only a 780-hp engine it carried the required armament of two 20-mm Type 99 (Oerlikon) cannon in the wings and two 7.7-mm (0.303-in Vickers Mk V) machine-guns above the nose, plus the large internal fuel capacity to meet the difficult range requirement of 1850 km (1150 miles), and still possessed outstanding power of manoeuvre (almost as good as the A5M).

During 1939 the fractionally larger Nakajima Sakae 14-cylinder two-row engine was substituted, and the new power of 925 hp dramatically improved performance still further. Though still less powerful than rival machines, it had markedly superior manoeuvrability. Coupled with its firepower, this made it a formidable opponent.

Two squadrons of production machines, designated A6M2 Model 11, and also known from the Japanese year 2600 (1940) as Type 0 or Zero-sen, embarked for China on July 21, 1940. In August they were in action high over Chungking, and on their first engagement destroyed all the Chinese (or American Volunteer Group) fighters they encountered so no word of the A6M reached their base. But within days the presence of this extremely deadly warplane was known to the Chinese and the AVG leader, the American General Claire Chennault, who cabled a detailed account to Washington. This vital document appears to have been ignored, so when the same Japanese fighter burst on the scene over Pearl Harbor on December 7, 1941 it was completely unknown to the defenders.

By this time there were well over 400 Zeros

Below and centre: **This A6M2 was captured by the Allies in the Aleutians and returned to the United States for testing.** ***Bottom:*** **An A6M3 assembled from the parts of five shot-down Zeros**

Associated Press

Associated Press

USAF

Fuji Photo

Zeros (foreground) and Aichi D3A2 'Kate' dive-bombers about to take off from a carrier

in service, mostly of the A6M2 Model 21 type with balanced ailerons with geared tabs linked to landing-gear position for better high-speed handling, and folding wing-tips. They appeared all over the gigantic expanse of the Pacific theatre, covering vast distances with drop tanks and easily disposing of the inferior Allied fighters they encountered.

The aircraft that had comparable firepower, such as the Hurricane, were sluggish and could not come near the A6M in rate of climb, turn radius or dogfight capability. Those that could almost stay with it in a fight, such as the biplanes and the Brewster Buffalo and Boeing P-26, lacked performance and firepower. As an all-rounder the early A6M was a masterpiece, and it did more than any other single weapon to give Japan the ascendancy and initiative throughout the first 18 months of the Pacific War in 1941-43.

In April 1942 production began of a floatplane version. This was first flown in February 1941 to provide air superiority during amphibious landings before the capture of land airstrips, and at times when carrier forces were not present. This machine, the A6M2-N, had a large central float and stabilizing outrigger floats on neat single struts. It was developed by Nakajima, and 327 were delivered before termination in September 1943. Code-named 'Rufe' by the Allies, the 2-N was probably underpowered, for its extra drag reduced speed to 438 km/h (272 mph) at optimum height and also impaired manoeuvrability; nevertheless, these fighter seaplanes were heavily engaged from Guadalcanal to the Aleutians and were respected by the Allies.

Another A6M version of 1942 was the A6M3, with the 1130-hp Sakae 21 engine and with the wing-tips removed. Though the latter change eliminated the aileron tabs the rate of roll actually improved; in other respects, save maximum speed, the Model 32 was a disappointment. The Allies thought the square-tipped fighter a new type and called it 'Hap' and then 'Hamp' until with closer acquaintance it finally became 'Zeke 32'.

Most Zero-sens were of the A6M5 type, with the short wing rounded off for better efficiency. The 5a (Model 52A) had a strengthened wing and extra 7.7-mm ammunition. The 5b (Model 52B) had extra protection and a 7.7 replaced by a 12.7-mm (0.5-in) gun, and the 5c had two 20-mm and two 13.2-mm Type 3 wing guns and a single 13.2-mm in the nose. Owing to the failure of Mitsubishi's planned Zero-replacement, the A7M Reppu, the A6M soldiered on long after it had become outperformed and outgunned. Up against the Hellcat and Corsair it was often destroyed by a half-second burst, and the quality of the Imperial Navy pilots also sharply declined as the experienced ones were replaced by raw recruits.

Even so, in the hands of a skilful pilot, the Zero could still display some of its old ability. Saburo Sakai, one of the greatest Japanese fighter aces, and one of the few to survive the war, tells in his book *Samurai* how he was attacked by no less than 15 Hellcats. Alone, unable to escape from the faster American fighters, he rolled, spiralled, looped and rolled again, constantly dodging the hail of lead spitting from 90 machine-guns.

The Americans stayed with him until, drawing them into range of the antiaircraft defences of Iwo Jima, he was able to escape. Miraculously, when he landed, there was not a single bullet hole in his aircraft.

The A6M5 (Model 52) retained the reduced wingspan of the A6M3, but the tips were rounded and individual exhaust stacks fitted to provide thrust augmentation, giving a higher maximum speed. The prototype was completed in August 1943 and deliveries began in March 1944

But by 1944 such flying ability was rare. Few Zeros were able to beat Hellcats in the air—and even while Sakai was standing off his adversaries almost half the fighters in his wing were being blasted from the sky.

Marginal improvement in performance came in late 1944 when the Sakae 31, rated at 1210 hp with water/methanol injection, became available after very protracted development. The resulting A6M6 also had self-sealing tanks and wing rocket rails. The A6M7 was a dive bomber, and the A6M8 at last had the 1560-hp Mitsubishi Kinsei engine, not because the navy had given in to Mitsubishi's long pleading but because the Sakae plant was unable to supply enough engines. Most of the survivors of the 10 937 Zero-sens finished as suicide attackers, reflecting the gradual but inexorable decline in Japan's mastery of the skies. Most of the Zeros were made by Nakajima, which delivered 6217 not including the 2-N floatplane. Small batches were made of the 2-K two-seat trainer, a total of 508 being produced.

Span: (A6M1, M2) 12.12 m (39 ft $4\frac{1}{2}$ in); (A6M3-M8) 11 m (36 ft 1 in) *Length:* 9.07 m (29 ft 9 in) *Gross weight:* (M1, M2) 2410 kg (5313 lb); (M5, M6) 2743 kg (6047 lb); (M8) 3150 kg (6944 lb) *Max speed:* (M1) 489 km/h (304 mph); (M2) 508 km/h (316 mph); (M3) 541 km/h (336 mph); (M5, M6) 557 km/h (346 mph); (M8) 573 km/h (356 mph)

Cutaway illustration of the A6M2 Model 21, showing structural, engine and armament detail. Note the folding wingtips, introduced on this model to facilitate carrier stowage. The model 21 Zero was the main type in service during the runaway Japanese victories in the Pacific

A-7 US Navy attack bomber See **Corsair II**

A7V

German tank. The A7V was the first German-designed tank to see battle. It was developed by a War Ministry Committee, the *Abteilung Verkehrswesen 7* (whence came the title of the tank) using a Holt tractor chassis as the starting point. The first prototype was demonstrated on May 14, 1917 and was approved for manufacture, but the shortage of steels and manufacturing facilities in Germany made production slow and difficult. The first Tank Detachment of five vehicles was formed in January 1918 and after hurried training took part in the St Quentin battle of March 21, 1918. A total of 15 had been put into service by the time the war ended and only 20 were ever produced, since German industry was committed already to the manufacture of more conventional weapons.

The A7V was armed with a captured Russian 57-mm 'Sokol' gun in the bows and six Maxim machine-guns distributed around the sides and rear. The driver and commander sat in the fixed turret, the remainder of the tank being occupied by a crew of 16 men. It was propelled by two Mercedes-Daimler engines of 150 hp each.

In 1918 an improved model, the A7V-U, was designed. This greatly resembled the contemporary British tanks in appearance, weighed 44 tons, had a seven-man crew and was armed with two 57-mm guns and four Maxims. Only a prototype was built before the war ended, but in many respects—notably suspension and armour—it was superior to contemporary British designs.

Length: 7.32 m (24 ft) *Width:* 3.05 m (10 ft) *Height:* 3.35 m (11 ft) *Weight:* 33 tons

A-9/A-10

German ballistic missile (project). Although it got no further than the blueprint stage at Peenemünde in 1941-42, this German project more than any other pointed the way to the multi-stage intercontinental ballistic missiles of postwar years. The scheme involved launching an improved A-4 rocket, fitted with dart-like wings and enlarged aerodynamic rudders, from a large liquid-fuel booster. Specified thrust of the booster, known as the A-10, was about 200 000 kg (441 000 lb).

Like the A-9 itself, the booster was to be stabilized by four steerable gyro-controlled graphite vanes located in the same plane as the aerodynamic surfaces around the exit of the rocket engine nozzle and impinging in the exhaust.

The nose section of the booster was slotted to receive the delta-winged A-9 which was backed onto a thrust ring. Overall length was 33.5 m (110 ft) and fuelled ready for takeoff the combination would have weighed over 100 tonnes.

The early part of the flight was to follow the practice of a conventional two-stage rocket and after the booster had separated the winged rocket was expected to reach a cut-off speed of more than 10 000 km/h (6214 mph). Having reached the zenith of the trajectory and begun to drop back into denser air, aerodynamic controls were to terminate the dive whereupon the remainder of the flight was to proceed in a series of aerodynamic 'skips', ending in a protracted glide. Assuming that launch sites were established in western France or Portugal there was the possibility of reaching coastal areas of the United States. However, in the pre-atomic age, a one-tonne warhead combined with poor target accuracy would have had little more than nuisance value.

A-9/A-10 two-stage rocket combination. One of the schemes for a two-stage transatlantic missile used the proposed A-9 guided missile, a winged development of the A-4, as the second stage

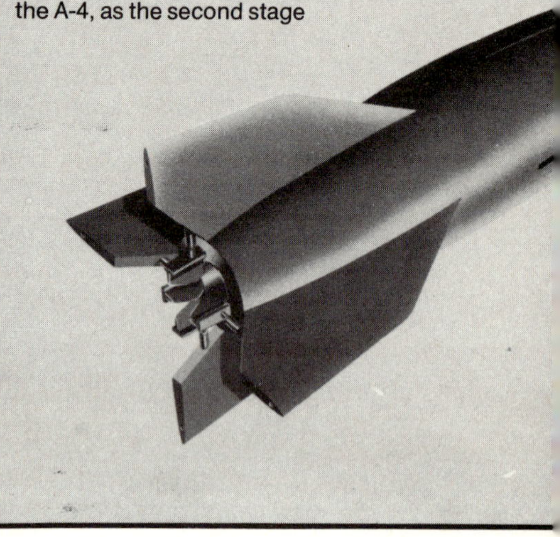

K W Gatland

Some of the 16-man crew of an A7V relax aboard their vehicle. Poor design and a production total of only 20 reflect the low priority accorded tanks by the German command

A-9/A-10

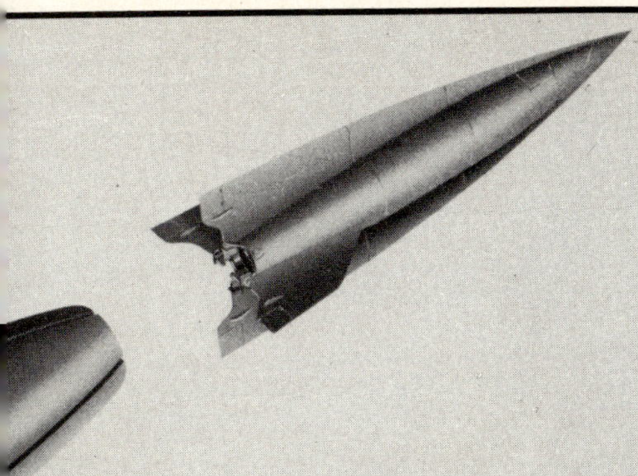

(A-9) Length: 14 m (46 ft) *Diameter:* 168 cm (66 in) *Thrust:* 25 000-27 000 kg (55 116-59 525 lb)×67 sec *Range:* 4000-4830 km (2485-3000 miles)
(A-10) Length: 20 m (65.5 ft) *Fin span:* 9 m (29.5 ft) *Diameter:* 4.1 m (13.5 ft) *Thrust:* 200 000 kg (441 000 lb)×approx 50 sec

***Above:* The A7V, a 30-ton metal box on wheels armed with a 57-mm gun and six Maxims. *Below:* 'Elfriede', the first German tank captured by the British, in early 1918**

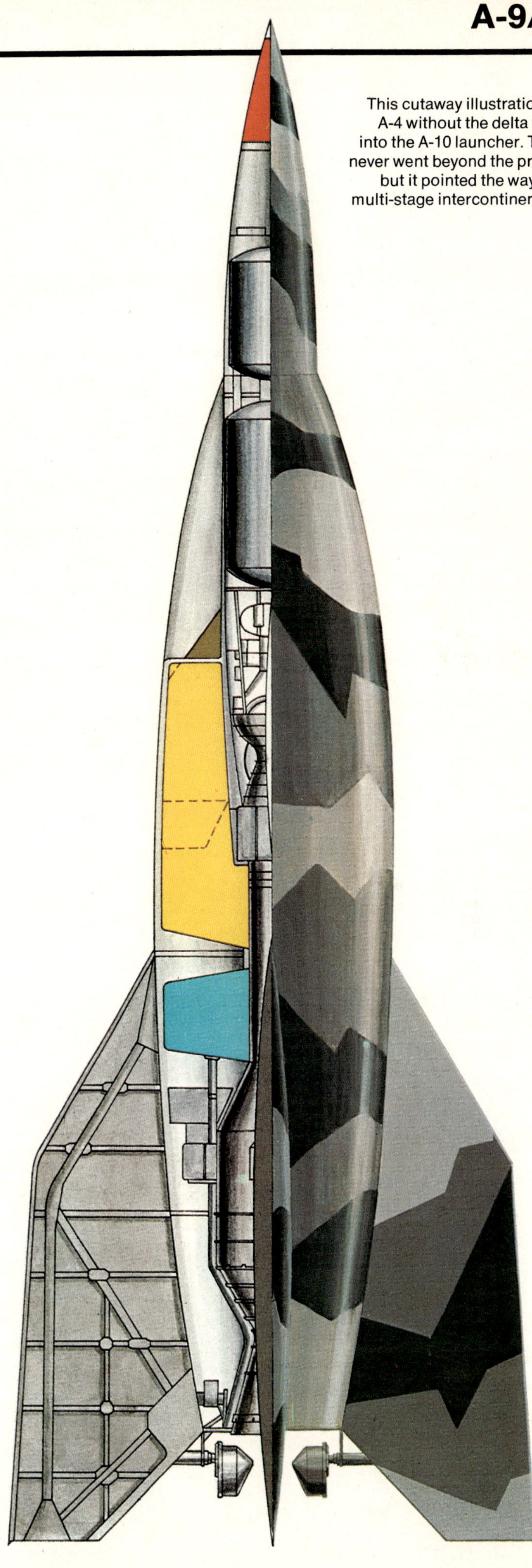

This cutaway illustration shows an A-4 without the delta wings fitted into the A-10 launcher. The weapon never went beyond the project stage, but it pointed the way to postwar multi-stage intercontinental missiles

A-10 Fairchild

The A-10's main weapon is the General Electric GAU-8 Avenger seven-barrelled 30-mm cannon. Weighing 1901 kg (4191 lb) with 1350 rounds of API (armour-piercing incendiary) ammunition, the total system is 6.4 m (21 ft) long, and fires at either 2000 or 4000 rds/min. As well as API ammunition it can fire high-explosive rounds at soft-skinned targets

USAF

USAF

A-10 Fairchild

USAF close-support attack aircraft. Unique among major warplanes, the Fairchild A-10A marks a trend away from performance (in terms of speed) in favour of weapon load, flight endurance, lethality and survivability. The first jet ever deployed specifically to fight in land battles, it promises to do efficiently the close air support (CAS) missions that have been a primary role in all recent wars for fighters and bombers that were actually quite unsuited to the task.

The A-10A is a fixed-wing unswept weapon platform powered by two 9065-lb (4119 kg) General Electric TF34-100 turbofans mounted high at the rear and giving no smoke and minimal infrared signature. The wing has a high-lift profile, down-turned tips and powerful flaps for gross-weight takeoffs from rough front-line strips never more than 1230 m (4000 ft) in length. Under the wing are pods for the main gears, which are interchangeable on each side like the twin tails and can land the aircraft even when retracted. From tip to tip are 11 stores pylons for up to 7257 kg (16 000 lb) of any tactical stores. Much of the nose is a titanium bath up to 3.8 cm (1.5 in) thick within which the pilot sits in an Escapac seat, while the biggest item in the whole aircraft is a tank-killing GAU-8 Avenger 30-mm gun firing shells able to pierce heavy armour at up to 4000 rds/min and fed by a drum holding 1350 of these impressive rounds. No radar is fitted, sensors and ECM (electronic counter measures) being carried in external pods.

Despite opposition from people unable to see that a 350-mph (563-km/h) warplane might be useful, the 'AX' (attack experimental) project survived years of scrutiny and eventually proved the viability of the non-VTOL (vertical takeoff and landing) fixed-

The Fairchild A-10, the first USAF aircraft developed specifically to provide aerial fire in support of ground troops

wing machine in front-line support of the foot soldier. In comparison with such machines as the A-1 Skyraider, F-100 and F-4 Phantom the A-10 is dramatically more effective and versatile, though due to inflation and reduced production-rate its price has escalated above the contract figure of $1.4 million (in 1970 dollars).

The first of two prototypes flew on May 10, 1972, and after a hard fly-off evaluation against the Northrop A-9A in California the A-10A was put into production for the USAF Tactical Air Command. The first TAC aircraft was accepted on March 20, 1976. In the second half of 1976 the 333rd Tactical Fighter Training Squadron—previously equipped in Thailand with the A-7, the A-10's closest rival—was working up at Davis-Monthan AFB, and by 1980 a 72-aircraft wing is to be deployed to Western Europe, the area most urgently threatened by large forces of potentially hostile armour. Fairchild is also hoping to find many other customers for this true 'battlewagon'. Even on the basis of planned USAF orders alone, for 733 aircraft, the A-10A will be about a $6000 million programme.

Span: 17.53 m (57 ft 6 in) *Length:* 16.56 m (54 ft 4 in) *Gross weight:* 21 400 kg (47 200 lb) *Max speed:* (clean) 713 km/h (443 mph); (with weapons or tanks and countermeasures pods) approx. 563 km/h (350 mph)

A-11 Original designation for SR-71/YF-12A US strategic reconnaissance aircraft
See **Blackbird**

USAF

Up to 7257 kg (16 000 lb) of stores can be carried on the A-10's ten underwing pylon stations, including both conventional and laser-guided weapons, rockets, cluster bombs and Maverick missiles, as well as ECM, chaff and flare pods. The aircraft above is carrying six Mavericks on two triple launchers, while that below is armed with a Hobos electro-optically guided bomb and a laser-guided bomb

USAF

The A-10's two General Electric TF34-100 turbofans were chosen for quietness, smokelessness and minimum infrared signature, and are positioned at the rear of the fuselage to minimize the risk of foreign-object ingestion when using rough front-line airstrips. The undercarriage is designed to land the aircraft even when retracted, and the entire airframe is hardened to withstand 23-mm shells

Structure, armament and engine details of the Fairchild A-10

A11 Aero

Czechoslovakian multi-role biplane. It is remarkable that in the years immediately after the First World War the new company of Aero (Aero Tovarna Letadel), in the newly created country of Czechoslovakia, should have been one of the most successful builders of combat aircraft in the world. Of all Aero's many designs the most significant was the A 11 of 1922. A classically simple single-bay biplane of mixed wood and steel-tube construction, the A 11 had every good quality, being efficient, robust, reliable, easy to maintain and a delight to fly. A two-seater, it was basically classed as general purpose, carrying bombs, two machine-guns, cameras and other gear, depending on the mission.

No fewer than 440 were built in 22 versions with a profusion of engines, the 240-hp Walter IV being chosen for the data below. Important variants were the Ab 11 bomber, A 11N night fighter and a Finnish variant with skis which later had a 400-hp Jupiter. The A 21 and 25 were trainers. From the A 11 stemmed such important machines as the A 30 reconnaissance bomber (again with many engines and variants) and the A 32 attack bomber of 1927-33 powered usually by a Jupiter.

Following the A 11 came a long series of successful military biplanes. In 1923-24 the Czech Air Force received 20 A 18 high-altitude fighters, with Walter W-IIIa engine and twin Vickers guns. In 1926-33 it received 79 bigger A 11 versions designated A 30, with 450/500 hp Skoda L (variants had Jupiter, Lorraine or Praga engines), plus 31 smaller A 32 close-support bombers with 450-hp Walter Jupiter IV radial; in all 116 A 32 variants were built by 1932, 16 being supplied to Finland. The final Aero biplanes were the completely new A 100 and A 101. In 1934-36 the Czech AF received 44 A 100 reconnaissance bombers, with 750 hp Avia Vr-36 V-12 and two fixed and two movable Mk 30 machine-guns as well as 29 A 101 bombers with 800-hp Praga (Isotta licence) Asso and the same 600-kg (1320-lb) bombload. In 1936-38 64 obsolescent Ab 101s were delivered, with 860-hp Praga (Hispano-licence) 12 Ydrs.

Span: 12.80 m (42 ft 0 in) *Length:* 8.11 m (26 ft 11 in) *Gross weight:* 1480 kg (3264 lb) *Max speed:* 214 km/h (133 mph)

A-17 USAAF attack bomber See **Nomad**

A-24 USAAF designation for US Navy SBD dive-bomber See **Dauntless**

A-25 USAAF designation for US Navy SB2C dive-bomber See **Helldiver**

A-26 USAAF/USAF attack bomber See **Invader**

A-28, -29 USAAF bomber See **Hudson**

A-30 USAAF bomber See **Baltimore**

A-31 USAAF dive-bomber See **Vengeance**

A-32 Swedish attack bomber See **Lansen**

A-35 USAAF dive-bomber See **Vengeance**

A-36 Attack bomber version of USAAF P-51 fighter See **Mustang**

A-37 US light attack aircraft See **Dragonfly**

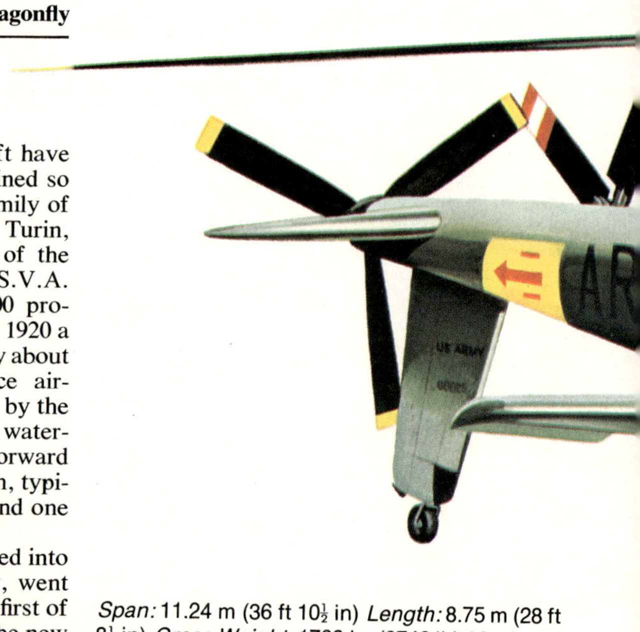

A.300 Ansaldo

Italian multi-role biplane. Few aircraft have been used in such numbers yet remained so little known as the Ansaldo A.300 family of general-purpose biplanes. Ansaldo, of Turin, had built the fastest fighting scouts of the First World War, the Balilla and S.V.A. After the Armistice the bigger A.300 prototype went on a sales tour in 1919. In 1920 a few two-seat A.300/2 were followed by about 90 three-seat A.300/3 reconnaissance aircraft, powered like nearly all versions by the excellent 300-hp Fiat A.12bis V-12 water-cooled engine. This was a straightforward two-bay biplane of mixed construction, typically armed with two fixed Vickers and one aimed from the rear cockpit.

In 1923 Ansaldo, soon to be absorbed into the Fiat empire and lose its identity, went into production with the A.300/4, the first of which had flown the previous year. The new Fascist government had created a separate air force, the Regia Aeronautica, and the Ansaldo biplane became its standard multi-purpose equipment; it served in Italy, Sicily, Sardinia, Corfu, Libya and even Eritrea, and was still in use in 1940. Total production is estimated at about 700, greater than the total for any single type of the 1920s except the Breguet XIX. It was similar to the A.300/3 but for the water-cooling radiator and details of equipment; some had only two seats. The A.300/4 was used as a reconnaissance aircraft, for surveying, bombing, ground attack, air fighting and communications. From 1929 most were advanced trainers. There were several unimportant later versions.

Span: 11.24 m (36 ft $10\frac{1}{2}$ in) *Length:* 8.75 m (28 ft $8\frac{1}{2}$ in) *Gross Weight:* 1700 kg (3748 lb) *Maximum speed:* 200 km/h (124 mph)

AA-1 Soviet air-to-air missile See **Alkali**

AA-2 Soviet air-to-air missile See **Atoll**

AA-3 Soviet air-to-air missile See **Anab**

AA-5 Soviet air-to-air missile See **Ash**

AA-6 Soviet air-to-air missile See **Acrid**

AA-7 Soviet air-to-air missile See **Apex**

AA-8 Soviet air-to-air missile See **Aphid**

No less than 22 different versions of the A 11 were produced by the Czech Aero company during the early 1920s. The standard A 11, illustrated, with the 240-hp Walter W-IV engine, was the principal model: others included the A 11N night reconnaissance version and Ab 11 and Ab 11N bombers with 240-hp Danek Perun II engines, as well as trainers and a floatplane

The AH-56A Cheyenne was developed by Lockheed for the AAFSS (Advanced Aerial Fire Support System) forerunner of the AAH competition. The US Army ordered 375 Cheyennes in 1968, but cancelled the contract the following year. Armed with either a 7.62-mm Mingun (shown) or a 40-mm grenade-launcher in the nose and a 30-mm cannon in a barbette below the rear fuselage, and able to carry Tow missiles or 2.75-in rockets, the aircraft was 16.66 m (54 ft 8 in) long with a rotor diameter of 15.62 m (51 ft 3 in), had a gross weight of 8301 kg (18 300 lb) and was capable of a maximum speed of 393 km/h (244 mph) at sea level

AABNCP US advanced airborne command post (modified Boeing 747) See **E-4**

AAH (Bell YAH-63A/ Hughes YAH-64A)

US Army helicopter programme. Since November 1972 the US Army, the world's largest user of helicopters, has been trying to purchase a helicopter designed from the start for battlefield combat duty, with the programme designation of AAH (Advanced Attack Helicopter). The AAH will eventually replace the Cobra family of gunships and make up for the abandonment of the extremely complicated and costly AH-56A Cheyenne which was tailored not only to the attack mission but also to reconnaissance, target designation and escort missions accompanying troop-carrying helicopters in missions of the kind then (1968) being flown in Vietnam. The AAH is smaller, cheaper, much more agile and more survivable, all these qualities having been specified in the US Army requirement, and all being necessary for its intended battlefield combat role.

In June 1973 Bell began building the YAH-63A, first flown on October 1, 1975, and Hughes the rival YAH-64A flown one day earlier. Both are tailored specifically to attack, making no pretence at any ability to lift troops or cargo (though having great potential payload). Though much smaller than the UTTAS transport helicopters they have the same power: two 1536-hp General Electric T700 turboshaft engines designed to be reliable in the harshest environments and to give no visible smoke and minimal infrared signature. The main and tail rotors of both are extremely advanced to allow violent nap-of-the-earth manoeuvres. The Bell has a main hub with elastomeric torsion springs, two broad blades and hand folding. The Hughes has four smaller blades with multiple load paths so that flight can be continued with parts shot through. Both helicopters are, as far as possible, resistant to 23-mm fire.

In both the most costly item is the complex stabilized sight system, for use by day or night with optics, laser and infrared, for target acquisition and weapon aiming. Bell seats the pilot in front (unlike its earlier Cobras), while Hughes puts the gunner in front. Bell uses the GE XM188 three-barrel 30-mm cannon, while Hughes uses the Hughes Aircraft 30-mm 'chain gun' in a pod on the left of the cockpit. Stub wings carry multiple pylons for 16 Tow or Hellfire laser-guided anti-tank missiles, 76 rockets or other weapons. In the Bell these wings also carry the main landing gears, while the Hughes has a tailwheel configuration with well-sprung main legs on each side of the armoured cabin.

The principal role of whichever AAH is chosen will be tank-killing. All battlefield operations in future wars will demand maximum use of natural cover and stealthy approach with minimal emissions—aural, visual, radar reflections or infrared—that would betray the helicopter's presence. The pilot will fly, navigate and man the remotely controlled gun, while the gunner will sight the other weapons using a head-down display (ie, he will look at his instrument display, not out of the window).

In December 1976 $317.7 million was voted for 50 months further development of the Hughes YAH-64, after which 536 AH-64s may be bought at a cost of $3580 million.

Rotor diameter: (B, Bell) 15.54 m (51 ft 0 in); (H, Hughes) 14.63 m (48 ft 0 in) *Overall length:* (B) 18.52 m (60 ft 9 in); (H) 17.40 m (57 ft 1 in) *Gross weight for primary mission:* (B) 6872 kg (15 150 lb); (H) 5987 kg (13 200 lb) *Max speed:* (both) approx. 306 km/h (190 mph)

Below and opposite: The YAH-63, Bell's entry in the US army's AAH competition

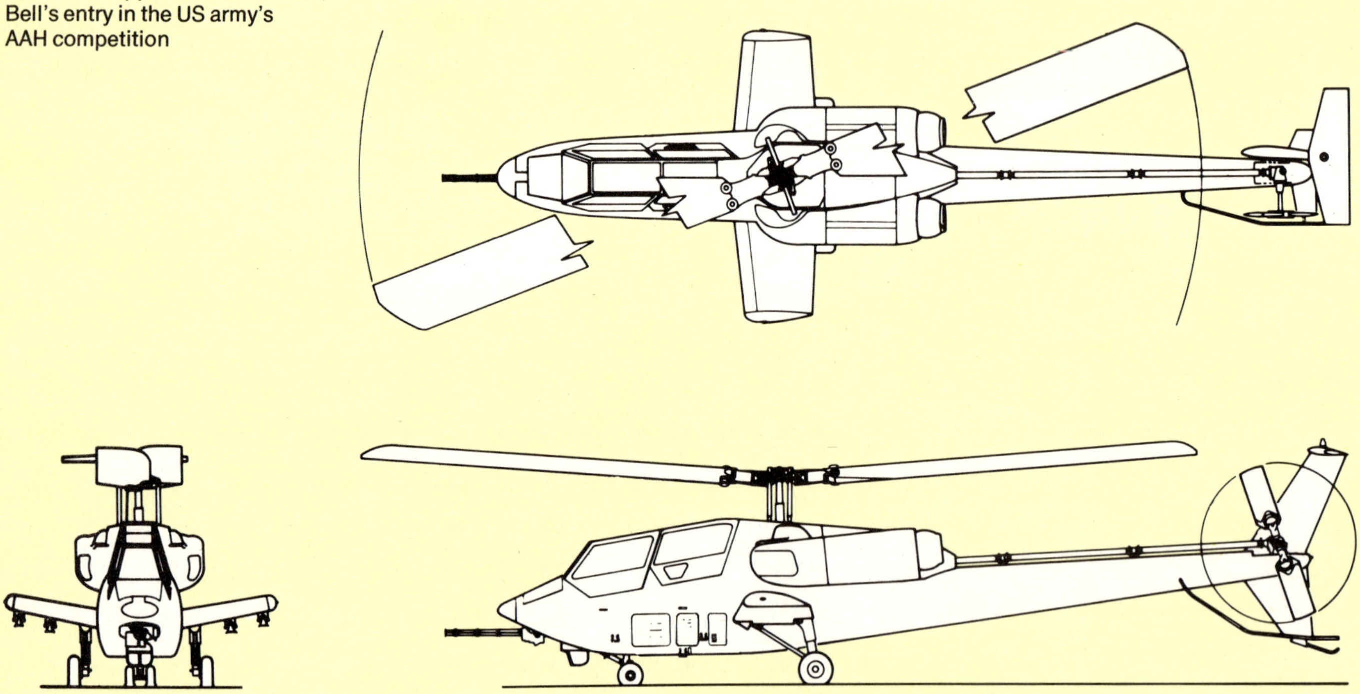

Bell Helicopters

The YAH-63 has a tricycle landing gear, with main gears mounted on the stub wings

Bell Helicopters

Bell Helicopters

The YAH-63 seats the pilot in front with the co-pilot/gunner behind. The pilot will be responsible for defending the machine with the 30-mm XM-188 multi-barrel cannon, sited in front of the aircraft to minimize muzzle blast damage, while the gunner sights missiles and rockets.

Bell Helicopters

Flapping axis moment springs on the YAH-63 rotor hub provide zero-g control, and enable it to be started in winds of up to 60 knots. The two-blade rotor configuration was chosen for simplicity and to minimize the risk of hits on the blades

Hughes Helicopters

The YAH-64 during test firing of 2.75-in rocket. As well as the Hughes Chain Gun it can carry up to 16 Hellfire guided missiles, 76 2.75-in rockets or combinations of the two

Hughes' AAH competitor, the YAH-64, retains the conventional helicopter gunship arrangement of pilot in the rear and co-pilot/gunner in the front of the cockpit. The Hughes Chain Gun, a three-barrelled 30-mm weapon designed specifically for the AAH and able to fire at any rate from single shots to 700 rds/min, is mounted in a turret under the fuselage

Hughes Helicopters

One of the main demands of the AAH competition was 'survivability': the four-blade main rotor of the YAH-64 is resistant to 23-mm hits, and was chosen for quietness and to minimize visual 'flicker perception'

Hughes Helicopters

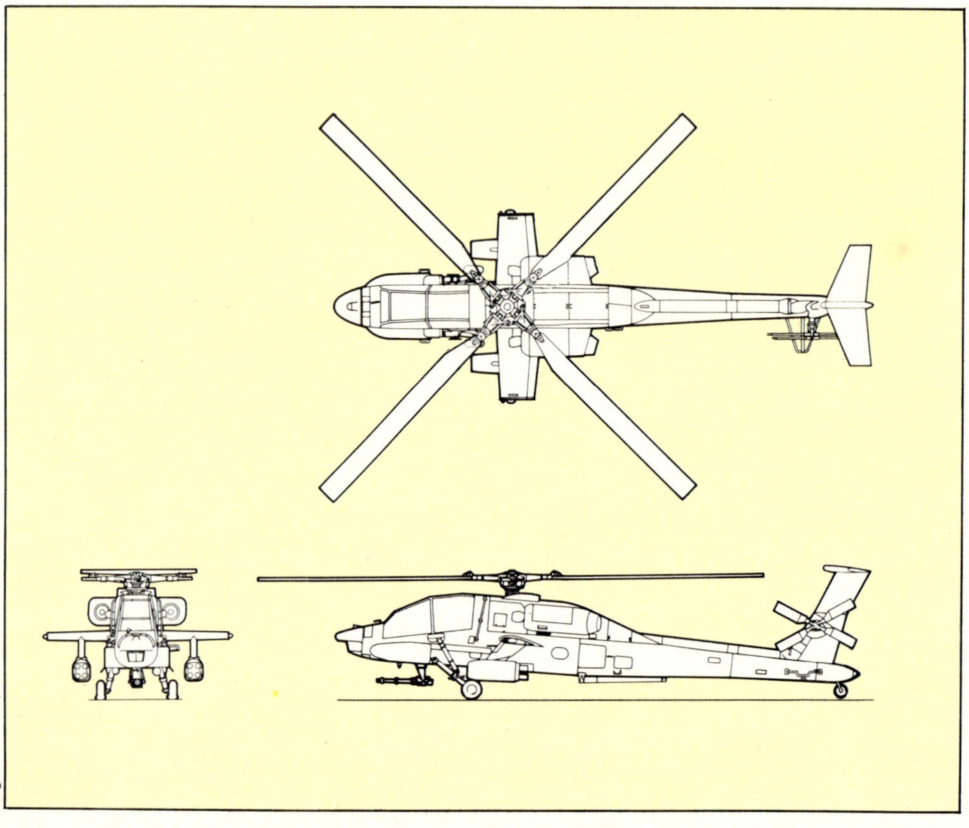

Hughes Helicopters

Among AAH requirements were flight performance and manoeuvrability to permit nap-of-the-earth flight and pop-up 'fire and evade' missions

Hughes Helicopters

AAM-1

Japanese air-to-air guided missile. Japan has been active in the field of solid-fuel air-to-air missiles for many years combining the talents of Mitsubishi and the Technical Research and Development Institution (TRDI) of the Japanese Defence Agency. The first operational example, known as the Mitsubishi AAM-1, was introduced in 1969. A pursuit-course, fair-weather missile it resembles the AIM-9 Sidewinder and has an effective range of about 5 km (3.1 miles). The infrared guidance system operates in conjunction with steerable cruciform canard fins.

AAM-1 replaced and supplemented AIM-9B and 9E variants of Sidewinder on F-86 and F-104J interceptors of the Japanese Air Self-Defence Force (JASDF) and was earmarked for the FST-2.

Length: 250 cm (98.4 in) *Diameter:* 15 cm (5.9 in) *Span:* approx. 50 cm (19.7 in) *Weight:* 76 kg (167.5 lb)

AAM-2

Japanese air-to-air guided missile. A more recent product of the TRDI/Mitsubishi partnership is the solid-fuel AAM-2 which supersedes early versions of the Sidewinder and AAM-1. The maximum effective range under clear-weather conditions is about 5 km (3.1 miles). Close attention has been paid to widening the angular range of interception by improved infrared guidance equipment and to limiting the effect of countermeasures, eg decoy flares. Control is applied via cruciform steerable rear fins.

Air-launched trials were scheduled for completion by late 1977 but it remained to be seen if the JASDF would select the weapon in competition with the improved Sidewinder AIM-9L as a replacement for Falcon AIM-4Ds on F-4EJ Phantoms. It was also expected to arm the FST-2.

Length: 220 cm (86.6 in) *Diameter:* 16 cm (6.3 in) *Span:* 49 cm (19.3 in) *Weight:* 74 kg (163 lb)

AAT 52

Arme Automatique Transformable Modèle 52, French general purpose machine-gun. Developed by MAS (Manufacture d'Armes de St-Etienne), and also known as the AA 52 and MAS 52, it is the latest in a long line of machine-guns which have shown rather less merit than their contemporaries. However, the AAT 52 is a reasonably effective design. In concept it is an amalgam of several other weapons, and is unusual among GPMGs in being operated by blowback, albeit with delay. The two-piece bolt is similar to that of the CETME rifle of 1946, and the front part has to be blown back first of all before the rear portion is unlocked, the two then move together against a spring operating with mechanical advantage.

The primary extraction of the case tends to be violent, as with all blowback actions, and the chamber is fluted to prevent cases sticking. Even so, a substantial number of cases expand and split on extraction, though this does not seem to lead to any trouble. It does, however, show that the system is working near its limit.

On the credit side, the gun is reasonably light and portable, though it has a long unsupported barrel which can be difficult to change when hot, and there is a suspicion that it droops with heat and alters the sight line. The mechanism is simple and easy to strip and maintain, and the belt feed works well. There is a stable and strong tripod to mount the gun in the medium role, and there are two types of barrel, a light and a heavy, for the light and sustained fire tasks. A folding butt offers some advantage in movement on foot, and there is a sensible carrying handle.

The AAT 52 has been in service since the mid-1950s and the French army has said that it will continue with it at least until 1990.

Weight unloaded: 9.88 kg (21 lb 12 oz) *Length:* 990 mm (39 in) *Operation:* Delayed blow-back *Calibre:* 7.62-mm NATO *Magazine:* Belt feed *Rate of fire:* 700 rpm *Muzzle velocity:* 823 m/sec (2700 ft/sec)

Abbot

British self-propelled gun. Development of the Abbot began in the mid-1950s with the acceptance of 105-mm as the standard NATO calibre for close support guns. The chassis was based on the FV432 APC design which was being developed concurrently, while the gun and its associated ammunition was developed by the Royal Armaments Research and Development Establishment. The first prototypes were built in 1961, testing began in the following year, and it was formally approved for issue in 1965.

On the vehicular side Abbot is powered by a Rolls-Royce K60 multi-fuel engine developing 240 bhp, and drives through an Allison TX200 six-speed automatic gearbox. Suspension is by five rubber-tyred wheels on each side, sprung by torsion bars and operating a manganese-steel track with rubber pads. Level road speed is about 48 km/h (30 mph), it can cross a 2-m (6.75-ft) trench, climb a 60-cm (2-ft) obstacle and surmount a 30° slope. It is also capable of swimming, using a permanently-fitted screen for flotation and driving by the paddle-wheel action of its tracks at about 3 knots. The interior of the vehicle can be sealed and pressurized for operation in a CBR environment.

The gun, officially known as the Ordnance 105-mm L13, is 30 calibres long and carries a fume extractor and muzzle brake. It can be elevated from −5° to +70° and, fitted in a turret, can be traversed through 360°. The breech mechanism is a semi-automatic vertical sliding block, and the cartridge is fired electrically.

The ammunition is in the form of separate cartridges and shells. The standard projectile is a 15.87-kg (35-lb) high-explosive shell, while smoke, illuminating and squash-head antitank shells are also provided. The brass cartridge case contains a five-part propelling charge which can be adjusted to provide the desired range and trajectory combination. There is also a 'Super' charge in a separate case, non-adjustable, which provides the maximum performance in terms of range and velocity. One notable feature is the gun's

The AAT-52 French general purpose machine-gun, a 7.62-mm delayed blowback weapon with a rate of fire of 700 rds/min

ability to fire the standard US and NATO 105-mm shells, using a special cartridge.

Maximum range with the high-explosive shell is 17 300 m (18 900 yards) and 38 complete rounds can be carried inside the turret for ready use; in normal field artillery positions, ammunition is supplied through the rear hatch from a Stalwart supply carrier. The normal rate of fire is six rounds per minute, but this can be doubled for short periods. A variant with two 30-mm guns has been tested.

Length overall: 5.84 m (19 ft 2 in) *Width:* 2.64 m (8 ft 8 in) *Height:* 2.44 m (8 ft 2 in) *Crew:* 4 *Range:* 483 km (300 miles) *Armour thickness:* 6-12 mm (0.24-0.47 in)

Vickers

The Abbot self-propelled 105-mm gun

A battery of Abbots on a firing range in West Germany. Standard NATO and US high-explosive, squash-head antitank, smoke and illuminating shells can be fired. The Abbot is fully amphibious, being fitted with a collapsible flotation screen

Sectional view of the Abbot showing the engine, suspension, traverse and elevation mechanism, ammunition and small arms stowage layout

IWM

IWM

Left: *The large mine decks and unusually high speed of the* ***Abdiel*** **class ships made them ideal for running supplies to Malta: here,** ***Welshman*** **unloads stores in July 1942.** ***Right:*** **HMS** ***Welshman*** **enters Grand Harbour, Malta, in 1942 disguised as a Vichy French destroyer with funnel caps and her sides painted to simulate a break in the deckline. The disguise twice succeeded in fooling enemy aircraft**

Abdiel

British fast minelayer class. These ships were among the most glamorous warships of the Second World War and also the most versatile. Although designed for the single purpose of laying mines in enemy waters, their high speed (for rapid transit to and from the laying area) and capacious decks made them useful for carrying stores and personnel. One of them spent her whole career carrying stores, and never laid a mine.

The most remarkable feature of the design was the provision of what amounted to a cruiser's horsepower in a hull little larger than that of a destroyer. The designed speed was 40 knots, and this, coupled with the reputation that the ships gained by their dramatic dashes through the Mediterranean to Malta, led to exaggerated accounts of their speed. In fact the ships were rather overweight, and in any case the forecasts of the engineers proved over-optimistic, so that none reached more than a fraction over 37 knots on trials, whereas speeds well in excess of 40 knots were claimed. Still, a sea speed of 35 knots or more was extremely creditable when it is remembered that destroyers were only capable of sea speeds of 30-31 knots. Another virtue of the design was the heavy antiaircraft armament, which gave the ships some chance of survival in the dangerous waters of the Mediterranean.

The first four ships—the *Abdiel, Manxman, Latona* and *Welshman*—were built under the 1938 Programme; all but one were war losses by 1943, but a repeat class of two—the *Apollo* and *Ariadne*—had been started in 1941. These ships differed from the earlier quartet in sacrificing a pair of 4-in AA guns in favour of a heavier and more efficient close-range armament. Both groups carried 156 mines on four sets of rails at main deck level, two of which extended as far as the forward superstructure. The mines were winched aft and discharged through stern doors.

The most outstanding achievements of the class were two runs made, one to lay mines in the Gulf of Leghorn and one to run desperately needed supplies to Malta, by *Manxman* and *Welshman* disguised as Vichy French destroyers. This was aided by the flush deck and three funnels, which were unique in British ships of that size; a dark patch simulated a break in the deck level and caps on the funnels changed the appearance entirely. The ruse succeeded in deceiving the enemy on both occasions when the ships were sighted by German aircraft.

Displacement: 2650 tons (standard), 4000 tons (full load) *Dimensions:* 127.41 m (418 ft) oa×12.19 m (40 ft) ×4.88 m (16 ft) max *Machinery:* 2-shaft geared steam turbines, 72 000 shp=37 knots *Armament: (Abdiel, Latona, Manxman, Welshman)* 6×4-in (100-mm) AA (3×2), 4×2-pdr pom-poms (1×4), 8×.5-in (13-mm) machine-guns (2×4) later replaced by 20-mm (0.79-in) AA; *(Apollo, Ariadne)* 4×4-in AA (2×2), 4×40-mm (1.6-in) Bofors AA (2×2), 12×20-mm AA (6×2), 156 mines *Crew:* 246

Abercrombie

British monitor class (1915). When Churchill, First Sea Lord, and Fisher, First Lord of the Admiralty, were planning their big new construction programme in November 1914, they received an offer of four twin 14-in gun turrets from America. Bethlehem Steel had contracted to provide the main armament of the Greek battlecruiser *Salamis,* then building in Germany, but the British blockade prevented delivery. The Admiralty could therefore mount these guns in a fleet of monitors, planned to steam close inshore to bombard enemy bases. Four monitors were immediately ordered, the *Abercrombie, Havelock, Raglan* and *Roberts.*

The first monitor was completed by Harland & Wolff in May 1915, named *Admiral Farragut,* but this name was soon changed to *M 1,* then *Abercrombie.* Owing to her very bluff lines, she could barely make seven knots instead of the designed ten. She was, therefore, towed to her first operational deployment—support of the British and French landings at Gallipoli. Arriving in July 1915, her big guns with their 18 300-m (20 000-yd) range proved useful at suppressing the Turkish gunfire from across the Dardanelles and supporting Allied offensives. She remained in close support until the withdrawal in January 1916. For the rest of the war, she supported Allied activities in the Aegean, including bombardments on the Salonika front. Paid off soon after the Armistice, she was finally scrapped in 1927.

***Left:* The 1915 British monitor *Abercrombie* was built to take one of the four US-built twin 14-in turrets originally ordered for the Greek cruiser *Salamis*. *Right:* The 1943 *Abercrombie* adds her 15-in guns to a night bombardment off the coast of Gela, Sicily, in July 1943**

Abercrombie	(ex-*General Abercrombie* 19.6.15, ex-*M 1* 31.5.15, ex-*Farragut* 12.14, ex-*Admiral Farragut* 11.14) built Harland & Wolff, Belfast
Havelock	(ex-*M 2*, ex-*General Grant*) built Harland & Wolff, Belfast
Raglan	(ex-*Lord Raglan*, ex-*M 3*, ex-*Robert E. Lee)* built by Harland & Wolff, Govan
Roberts	(ex-*Earl Roberts*, ex-*M 4*, ex-*Stonewall Jackson*) built by Swan Hunter, Wallsend

Havelock was sold in 1927, and *Roberts* in 1936 after long service as a drillship. The *Raglan* was sunk with the small monitor *M 28* by the *Goeben* and *Breslau* off Imbros on January 20, 1918.

Displacement: 6150 tons (normal) *Length:* 116.6 m (335 ft) oa *Beam:* 27.43 m (90 ft) *Draught:* 3.05 m (10 ft) *Protection:* 50 mm (2 in) deck; 100 mm (4 in) sides; 254 mm (10 in) turret (main hull protected against mines and torpedoes by underwater bulges) *Armament:* 2 14-in (355-mm); 2 12-pdr; 1 3-pdr; 1 2-pdr

Abercrombie

British monitor (1943). Britain had built no monitors between the wars, so when HMS *Terror* was lost off North Africa in February 1941, only two were left. A new monitor was therefore ordered from Vickers-Armstrong, to be named *Abercrombie*. She was similar to the nearly completed *Roberts* and was fitted with the twin 15-in turret built in 1917 as a spare for the battlecruiser *Furious*, should the latter's 18-in single mountings prove unsuccessful. Strong resistance to air attack was necessary operating close to enemy shores, so she was equipped with eight 4-in, 16 2-pdr and 20 20-mm AA guns, first class radar and 4-in deck armour, plus 5-in side armour and bulges.

HMS *Abercrombie* was completed in May 1943 in time for the Allied landings in Sicily in July, when the 29 250-m (32 000-yd) range of her guns was put to good use supporting the Americans. On September 9 she started bombarding at the Salerno landings, but had the misfortune to strike a mine, which put her out of action for 11 months. After repairs she was exercising off Malta when she had the bad luck to hit another mine. This time she was out of action until July 1945, when she set off for the Indian Ocean to support operations against the Japanese in Malaya. The war was over before she could see any action; she was scrapped in 1954.

Displacement: 7850 tons (standard), 9720 tons (full load) *Dimensions:* 113.77 m (373 ft 3 in) oa×27.13 m (89 ft)×4.27 m (14 ft) max *Machinery:* 2-shaft geared turbines, 4800 shp=12 knots *Armament:* 2 380-mm (15-in) (1×2); 8×4-in (100-mm) AA (4×2); 16 2-pdr AA (1×8, 2×4); 20 20-mm (0.79-in) AA (20×1) *Protection:* see above *Crew:* 350

HMS *Abdiel* in early 1943 after her final refit. The original quad 0.5-in machine-gun mounts were replaced by seven single 20-mm Oerlikons in 1942. She is shown here painted in a 1943 Admiralty Disruptive camouflage scheme. The vessel was sunk by a mine in September 1943 while landing troops at Taranto

ABM-1 Soviet missile See **Galosh**

AC-3 Designation for Ansaldo-built Dewoitine D9 See **D9 Dewoitine**

AC-47 USAF gunship See **Dakota**

AC-119 USAF gunship See **Flying Boxcar**

AC-130 USAF gunship See **Hercules**

HMS *Shark*, one of the *Acasta* or 'K' Class destroyers built under the 1911-12 Programme

Acasta

British destroyer class (1911-23). The 20 vessels of this class were ordered under the 1911-12 Programme and completed during 1912 and 1913. Compared with earlier British destroyers they were designed for a higher speed and improved armament, mainly to counter German designs with reported speeds of between 30 and 33 knots. To attain the necessary speed the installed power was increased by over 50% giving them shaft horsepower equal to some of the battleships of the Fleet. Twelve of the class were to a standard Admiralty design while the remaining eight were 'specials' designed by the builders to Admiralty requirements. The standard ships were designed for a speed of 29 knots, and were virtually identical except that the three built by John Brown had Brown-Curtis instead of Parsons turbines.

The 'specials' differed from the standard design as follows:

Five Thornycroft ships were designed for a speed of 31 knots, which none achieved; the average speed on trial was 30.8 knots. The *Hardy* of this group was designed for a speed of 32 knots but only made 31.9 knots on trial. Arrangements were made to fit this ship with diesel engines but the machinery was not available when she was completed and so was never fitted.

Garland had the same appearance as the standard ships but was equipped with semi-geared turbines to give a designed speed of 30 knots. She made just over this speed on trial with 26670 shaft horsepower.

Fortune had a clipper bow and three funnels of equal height. She was designed for a speed of 30.5 knots and achieved 30.7 on trials with 27890 shaft horsepower.

Ardent was the first destroyer with a hull built on the longitudinal framing system. She had two funnels of equal height and was designed for a speed of 29.5 knots which she just achieved on trial.

On completion they formed the 4th Destroyer Flotilla, led by the *Swift* and from 1914 to 1916 served with the Grand Fleet. In 1916 they were transferred to the south coast, being based at Devonport or Portsmouth. The *Porpoise, Paragon, Unity, Victor* and *Ambuscade* joined the 6th Flotilla, Dover Patrol, in November and December 1916, but rejoined the 4th Flotilla at the end of 1917.

The first casualty of the class was *Lynx* which struck a mine and sank off the Moray Firth on August 9, 1915. Of the others, all but *Cockatrice, Paragon* and *Victor* took part in the Battle of Jutland. In daylight, the *Acasta, Shark* and *Christopher,* which were screening the 3rd Battle Cruiser Squadron, were engaged with enemy destroyers and light cruisers. *Shark* fired a torpedo at a light cruiser without success and was soon disabled by enemy gunfire. With only one gun left in action she drove back two enemy destroyers but was eventually torpedoed; she sank with her flag still flying. Her commander, Loftus W Jones, was awarded the Victoria Cross. The *Acasta,* also damaged in this action, later sighted and made an unsuccessful torpedo attack on the German battlecruiser *Lutzow*; she was under fire for 20 minutes and was eventually disabled. She was later taken in tow by the destroyer *Nonsuch* and arrived at Aberdeen two days later.

The main body of the 4th Flotilla was not heavily engaged during daylight but at night

Led by *Swift* the 4th Destroyer Flotilla was attached to the Grand Fleet in August 1914

HMS *Achates* of the 1927 *Acasta* Class as modified for convoy escort. A and Y guns have been removed to make way for a Hedgehog antisubmarine launcher forward and extra depth charge aft; a 20-mm Oerlikon has been added to each side of the bridge; and the 2-pdrs between the funnels have been replaced by 20-mm Oerlikons

they ran directly into the High Seas Fleet. The *Spitfire* actually collided with the battleship *Nassau*. Her bridge and masts were flattened by the blast of the battleship's guns (which fortunately could not depress sufficiently to actually hit her) and about 60 ft of her forecastle plating was torn away by the collision. She returned to the Tyne under her own steam with about 23 ft of the battleship's side plating wedged in her mess deck. In the confusion which followed the leader *Broke* collided with the *Sparrowhawk*, and while the *Broke* was endeavouring to extract herself the *Contest* hit and cut off five feet of the *Sparrowhawk*'s stern. The *Sparrowhawk* had to be sunk on the following day after an unsuccessful attempt to tow her home. (During this same period the German cruiser *Elbing* was rammed and sunk by the battleship *Posen* and the cruiser *Rostock* was hit by a torpedo from one of the 4th Flotilla, sinking later that night). The remainder of the Flotilla reformed and continued on its course but ran into the enemy fleet again half an hour later. During this second engagement the *Fortune* and *Ardent* were sunk and the *Porpoise* was disabled by gunfire. The rest of the Flotilla retired to the eastward and did not sight the enemy again.

Two more of the class were lost during the war; the *Paragon* was torpedoed and sunk off Calais during a raid by enemy destroyers on the night of March 17/18, 1917 and the *Contest* was torpedoed and sunk by a submarine in the English Channel on September 18, 1917.

Porpoise was sold in 1920 and eventually went to Brazil. The remainder were sold for scrap beteen 1921 and 1923.

Displacement: 935 tons (standard); 908-964 tons (special ships) *Length:* 83.7 m (276 ft 6 in) *Beam:* 8.37 m (27 ft 6 in) *Draught:* 2.74 m (9 ft) *Machinery:* 2-shaft turbines, 24 500 shp=29 knots *Armament:* 3 4-in (101-mm); 2 21-in (53-cm) torpedo tubes *Crew:* 75-77

Wright and Logan

HMS *Active* of the second *Acasta* or 'H' Class was built by Hawthorn Leslie and launched in 1929

Acasta

British destroyer class (1927-48). Constructed under the 1927 Programme, these destroyers were the first complete flotilla to be ordered after the First World War. Their design was based on that of two experimental destroyers, the *Amazon* and *Ambuscade*, built under the 1924-25 Programme. They in turn formed the basis for a large group ordered at the rate of one flotilla a year until 1936 and known collectively as the A to I Classes. They were the first destroyers in the Royal Navy designed to carry quadruple torpedo tube mountings. The leader was somewhat larger with additional staff accommodation and one more 4.7-inch gun, but was otherwise similar to the others.

The *Acheron* was fitted with a special Thornycroft boiler installation which operated with, for the time, a very high steam pressure of 500 lb per square inch; the boilers of the remainder of the class operated at 300 lb/sq in. All exceeded their designed speed on trials, the best performance being from *Anthony* which achieved 36.3 knots.

The Acasta Class of destroyers was very active during the Second World War and only three of the class survived the conflict. On June 8, 1940 the *Acasta* and *Ardent* were escorting the aircraft carrier *Glorious* from Norway to Scapa Flow when they were sighted by the German battlecruisers *Scharnhorst* and *Gneisenau*. The battlecruisers opened fire on *Glorious* at 14.30 at a range of 24 700 m (27 000 yd). The destroyers laid smoke to screen the carrier and endeavoured to carry out a torpedo attack. The *Ardent* fired eight torpedoes without success before sinking at 17.28. The *Acasta* was more successful; with her first salvo of four torpedoes she scored one hit on *Scharnhorst*, causing extensive damage. When *Acasta* came back through the smoke to make a second attack she was hit in the engine room by an 11-in shell and stopped. She launched the rest of her torpedoes but had to be abandoned. There were three survivors from *Ardent* and one from *Acasta*.

Acheron was mined and sunk during the night of November 17, 1940 while running post-refit trials south of the Isle of Wight.

In 1941 the remaining ships of the class were refitted, having the aftermost 4.7-in gun removed to make room for an increase in the antisubmarine armament to 70 depth-charges and four depth-charge throwers. In addition the after set of torpedo tubes was removed and all except *Achates* had a 3-in AA gun

fitted in its place. *Achates* also had A gun replaced by a Hedgehog A/S weapon.

In September 1941 the *Achates* hit a mine off the Norwegian coast. Her bow from the bridge forward was blown off but she was towed home and repaired by Swan Hunter between September 1941 and April 1942. On December 31, 1942 she was escorting the Russian convoy JW51B when it was attacked by the German cruiser *Hipper*. The *Achates* was hit twice by the German ship and she eventually capsized and sank.

The *Arrow* was badly damaged at Algiers on August 4, 1943 when a merchant vessel blew up in the harbour. She was written off as a total loss. The remaining three vessels, *Active, Anthony* and *Antelope,* were re-armed as antisubmarine escorts in 1944. This involved among other details the removal of A and X 4.7-in guns and the fitting of two 6-pdr Hotchkiss guns. These three were sold for scrap between 1946 and 1948.

Displacement: 1350 tons *Length:* 98.2 m (323 ft) *Beam:* 9.82 m (32 ft 3 in) *Draught:* 3.65 m (12 ft) *Machinery:* 2-shaft turbines, 34 000 shp=35 knots *Armament:* 4 4.7-in (120-mm); 2 2-pdr pom poms; 4 .303-in AA; 8 21-in (53-cm) torpedo tubes (2×4) *Crew:* 185

Acciaio

Italian submarine class. This class of 16 medium submarines was laid down in 1940-41 and completed in 1941-42.

Three were sunk by British submarines, four by Allied surface forces and one by aircraft. The *Volframio* was scuttled at La Spezia in September 1943, refloated by the Germans and sunk again by bombing in 1944. The *Bronzo* was captured by the British minesweepers *Boston, Cromarty, Poole* and *Seaham* off Syracuse on July 12, 1943; renamed *P 714* she was transferred to the French navy as the *Narval* and scrapped in 1948. The *Platino* was scrapped in 1948 and the *Nichelio* became the Russian *Z 14* in 1949. The *Giada* was converted to an electricity generating hulk to comply with the Peace Treaty, and renamed *PV 2*, but when Italy joined NATO she was rebuilt as a training submarine, and served until 1966.

Displacement: 701-714/860-871 tons (surfaced/submerged) *Length:* 60.8 m (199 ft $5\frac{3}{4}$ in) oa *Beam:* 6.44 m (21 ft $1\frac{1}{2}$ in) *Draught:* 4.78 m (15 ft 8 in) *Machinery:* (surfaced) 2-shaft diesels, 1400-1500 shp=15 knots; (submerged) electric motors, 800 shp=6.7 knots *Armament:* 1 100-mm (3.9-in); 1 or 2 20-mm AA; 2-4 13.2-mm AA; 6 21-in torpedo tubes (stern tubes omitted on *Argento, Bronzo* and *Volframio*) *Crew:* 45-50

Acheron

British destroyer class. Constructed under the 1910-11 Programme, the ships of this class were repeat editions of the *Acorn* Class except that they carried three boilers instead of four and two funnels instead of three. Fourteen were constructed to standard Admiralty specification with a designed shp of 13 500 giving a speed of 27 knots. Concern over the high speeds being achieved by new foreign destroyers, particularly those of Germany, led the Admiralty to give certain builders a free hand with the machinery installation of a further six of the class. These were *Archer* and *Attack* (16 000 shp giving 28 knots), built by Yarrow, the two contracted to Parsons (16 000 shp giving 28 knots) and the two Thornycroft ships (15 500 shp giving 29 knots). In addition Yarrow were allowed to build three 'Specials' to their own version of the *Acheron* design—the *Firedrake, Lurcher* and *Oak* (20 000 shp giving 32 knots). All were launched in 1911 except *Hydra* and the Yarrow Specials, launched in 1912.

Throughout the war, *Firedrake* and *Lurcher* operated with the Harwich force and the *Oak* served as a tender to the Grand Fleet Flagship. The remainder formed the 1st Destroyer Flotilla of the Grand Fleet until late 1916 when they were transferred to Channel or east coast bases. During 1917 and 1918 they all moved to the Mediterranean with the exception of *Ariel, Ferret* and *Sandfly* which were converted to minelayers and operated with the 20th Flotilla, in the North Sea. The *Ferret* was equipped to carry magnetic mines. Three of the class were lost—the *Attack* was mined on December 30, 1917 off Alexandria, the *Phoenix* was torpedoed by a submarine in the Adriatic on May 14, 1918 and the *Ariel* was mined on August 8, 1918 in the North Sea. The remaining ships of the class were sold for scrap between 1920 and 1922.

Displacement: 745-810 tons *Length:* 74.9 m (246 ft) average oa; 79.85 m (262 ft) Yarrow Specials oa *Beam:* 7.82 m (25 ft 8 in) *Draught:* 2.74 m (9 ft) *Machinery:* 2- or 3-shaft steam turbines, 13 500-20 000 shp=27-32 knots *Armament:* 2 4-in (101-mm) (2×1); 2 12-pdr (2×1); 2 21-in (53-cm) torpedo tubes (2×1) *Crew:* 72

Achilles

British self-propelled gun. The name Achilles was given to American-built M10 tank destroyers fitted with 17-pdr guns for British service use. The M10 was a self-propelled antitank gun derived from the chassis of the M4A2 Sherman tank, and was originally fitted with the American 3-in antitank gun M7. In late 1944 those in British service (which had been given the name Wolverine) had the 3-in gun removed and replaced by the 17-pdr gun Mk 5. The M10, which was diesel-engined, became the Achilles Mk 1, while the M10A1, petrol-engined, became the Achilles Mk 2. Although an official name, it was rarely used by the troops, who simply called them 'Seventeen-pounder M10s', irrespective of mark or model. Fast and hard-hitting, they were highly effective and well-liked by their crews, and they remained in service for some years after the war.

Length overall: 5.97 m (19 ft 7 in) *Width:* 3.05 m (10 ft) *Height:* 2.82 m (8 ft $1\frac{1}{2}$ in) *Crew:* 5 *Range:* 320 km (200 miles) *Armour:* 51 mm (2 in)

ACL-APX 80

French antitank system. This is an 80-mm recoilless rifle known as the Arme collective légère (or ACL) or light multi-purpose weapon and developed by the Atelier de Construction de Puteaux (APX). The rifle fires a fin-stabilized hollow-charge antitank projectile weighing 3.63 kg (8 lb), or a high-explosive antipersonnel projectile of 3.49 kg (7.7 lb) to a maximum range of about 800 m (2625 ft) for antitank and 1600 m (5250 ft) for antipersonnel.

The projectile is a gun-launched rocket. When loaded, it is enclosed at its rear end by a cartridge case with a Laval convergent-divergent nozzle which, held in the breach, acts as a recoilless venturi. The propelling charge in this cartridge case gives the projectile a muzzle velocity of about 400 m/sec (1300 ft/sec). Once the projectile is clear of the muzzle, the rocket motor is ignited and delivers thrust to accelerate to about 550 m/sec (1800 ft/sec). The fins, concealed within the cartridge case when loaded, are flung out after leaving the muzzle by centrifugal force, and are offset so as to maintain roll stabilization throughout flight.

Launcher length: 145 cm (57 in) *Weight:* 16.7 kg (36.8 lb) loaded, with sight *Armour penetration:* 120 mm (4.72 in)

RAC Tank Museum

The American M10 tank destroyer adapted for British service and known as the Achilles

Acorn

British destroyer class. The 20 vessels of this class were constructed under the 1909-10 Programme to an Admiralty design and completed during 1910 and 1911. Their design followed the general pattern of the time, but they were considered as an improvement on the destroyers of the previous year's Programme (the *Beagle* Class) because oil-fired boilers were again adopted. These had first been employed in the destroyers of the 1905-06 Programme but in the *Beagle* Class coal-fired boilers were installed owing mainly to concern over the possible restrictions of oil supplies in wartime. The advantages of oil were, however, considerable, giving a saving in machinery weight and engine-room complement, and allowing for improved engine performance. In the *Acorn* Class these advantages were used to reduce the size and cost of the ships. Compared with the *Beagle*s they had one less boiler and were some 200 tons lighter, while they carried a slightly heavier armament and had the same designed speed. All were fitted with Parsons turbines driving three screws, except *Brisk* which was fitted with a Brown-Curtis twin-screw turbine arrangement as an experiment. On trials all exceeded their designed speed, the best ship of the class being *Ruby* which averaged 29.4 knots with a little over 16 000 shaft horsepower. In general they were slightly faster than the *Beagle*s and could maintain high speeds for longer periods.

They carried 240 rounds of ammunition for the 4-in guns and 200 rounds for the two 12-pdr (3-in) guns. The torpedoes, which weighed nearly 1½ tons and carried a 127-kg (280-lb) charge, had a range of about 9144 m (10 000 yd) at 30 knots and 1828 m (2000 yd) at 50 knots. The ships handled and turned well, but when first completed suffered from heavy weather damage and all had their hulls strengthened prior to the outbreak of the First World War. In October 1913 they were redesignated as the H Class.

PPL

HMS *Nymphe* of the *Acorn* Class was repaired after having her stern blown off

The steam pipes on the centre funnel of the three vessels constructed by White *(Redpole, Rifleman* and *Ruby)* were more pronounced than in the other ships of the class, but otherwise there were practically no differences. Towards the end of the war most of the class were fitted with a 3-pdr antiaircraft gun and an antisubmarine armament of depth-charge racks and throwers.

On the outbreak of war they formed the 2nd Destroyer Flotilla of the Home Fleet based at Scapa Flow. As new destroyers joined the fleet from 1915 to 1917 they were gradually transferred to other theatres, partly because they had insufficient radius of action to operate efficiently with the Grand Fleet. Some of the class joined the 8th Flotilla based on the East coast but the majority went to the Mediterranean, where radius of action was less important. By 1918 all the ships in service were stationed in the Mediterranean.

The first casualty of the class was the *Goldfinch* which, in a fog on the night of February 18/19, 1915, was wrecked on Start Point, Sanday Island, on the northeast side of the Orkneys. The *Staunch* and *Fury*, which were among the first of the class to go to the Mediterranean, assisted in evacuating troops from Gallipoli in December 1915. Three of the class fell victim to submarines: the *Staunch* being torpedoed and sunk off Gaza, Palestine, on November 11, 1917; the *Comet* torpedoed and sunk in the Mediterranean on August 6, 1918; and the *Nymphe* having her stern blown off by a torpedo in 1917: however she was recommissioned the next year.

In June 1917 the *Minstrel* and *Nemesis* were lent to the Japanese, being commissioned in September as *Sendan* and *Kanran* respectively. The surviving units of the class were sold for scrap between 1920 and 1922.

Displacement: 720-780 tons *Length:* 74.98 m (246 ft) oa *Beam:* 7.77 m (25 ft 6 in) *Draught:* 2.58 m (8 ft 6 in) *Machinery:* 3-shaft turbines, 13 500 shp=27 knots *Armament:* 2 4-in (102-mm); 2 12-pdrs; 2 21-in (53-cm) torpedo tubes *Crew:* 72

HMS *Goshawk*, one of the *Acheron* or 'I' Class, a repeat of the *Acorn* Class, at sea with the 1st Destroyer Flotilla

PPL

The light cruiser *Fearless*, of the *Active* Class. *Fearless* led the 1st Destroyer Flotilla from 1914-1916, including the Battle of the Heligoland Bight in August 1914—the first full-scale naval battle of the war

IWM

ACRA

French gun-launched antitank weapon. Named from the acronym of Anti-Char Rapide Autopropulsé, ACRA was developed in France by the Groupement Industriel des Armements Terrestres (GIAT) at l'Atelier de Construction de Puteaux. The system combines the attributes of the high-velocity shell and a rocket missile with a high-precision laser aiming device. Two types of hybrid missile have been developed, both of 142-mm calibre: the larger version is guided and the smaller spin-stablized.

The guided version takes the form of a cylindrical rocket projectile with flip-out folding tail fins and four small anti-roll stabilizing fins. It is assembled into a cartridge case which is automatically ejected at the time of firing.

In firing trials from the gun of a battle tank (completed at the end of 1974), consistent hits were scored on stationary and moving targets up to distances of at least 3800 m (12 470 ft). The projectile, which leaves the gun muzzle at 150 m/sec (492 ft/sec), is boosted to a speed of 1800 km/h (1118 mph) by a solid-fuel sustainer motor. It flies along the laser beam directed at the target under the control of infrared guidance equipment acting on the steerable tail surfaces. The warhead has a shaped charge.

The smaller ACRA unguided rocket projectile is spin-stabilized for use against soft targets. After leaving the gun muzzle at 1980 km/h (1230 mph), a small sustainer motor boosts the speed to about 2520 km/h (1566 mph). Six tail fins flip out at the back to stabilize the missile in flight. Again, the standard of accuracy is high.

A steel 2-kg (4.4-lb) RDX-TNT warhead is specified for the guided version used against armoured vehicle targets. Alternatively, a cast-iron 2-kg (4.4-lb) TNT anti-personnel warhead is fitted to the unguided rocket.

Guided version: *Length with cartridge case:* 1.25 m (4.1 ft) *Weight:* 26 kg (57.3 lb) *Range:* 25-3800 m+(82-12 470 ft+) Unguided version: *Length:* 64 cm (25.2 in); 90 cm (29.5 in) with cartridge case *Weight:* 15 kg (33 lb); 21 kg (46.3 lb) with cartridge case

ACRA, the antitank missile developed for the French army by GIAT, shown in its laser-guided fin-stabilized form. Both this and the unguided spin-stabilized version are fired from conventional 142-mm (5.5-in) gun barrels

Acrid (AA-6)

Soviet air-to-air missile. Four of these large missiles are carried by the MiG-25 Foxbat A on underwing rails. Two versions of the missile have been produced, one having infrared and the other semi-active radar homing. Foxbat carries two of the IR variety inboard and two of the SARs outboard, the latter possibly operating in conjunction with continuous-wave target illuminating radars. It is likely that the weapons are ripple-fired in pairs, the IR round preceding the radar missile by about one second, as is the case with other Soviet air-to-air missiles.

Acrid employs a high-impulse solid-propellant rocket motor and has a speed of about Mach 3.5-4. It is clearly intended to engage targets at long range substantially higher or lower than the launch aircraft and it is likely that the radar detection range exceeds 80 km (50 miles) with the ability to 'lock-on' at distances up to 50 km (31 miles). Control is applied by wing ailerons and canard steerable fins which confer high manoeuvrability in the terminal stage of attack. The weapon is large enough to carry a 100-kg (220-lb) HE warhead.

Length: 580 cm (228 in) IR version; 615 cm (242 in) SAR version *Diameter:* 30-40 cm (11.8-15.7 in) *Span:* 225 cm (89 in) *Weight:* 700-850 kg (1543-1874 lb) *Estimated effective range:* 20 km (12.4 miles) IR version; 45 km (28 miles) SAR version

Active

British scout cruiser class. This class of three ships was an improved version of the *Boadicea* Class of 1908. Rated as scouts until 1913, when they became light cruisers, they were intended purely for reconnaissance and work with destroyers, and so sacrificed endurance and gunpower for speed. Even armour was restricted to a light protective deck, but in spite of this they gave good service in the North Sea during 1914-18.

Amphion was the first casualty of the First World War. While leading the 3rd Destroyer Flotilla in the destruction of the German minelayer *Konigin Luise* on August 6, 1914,

she struck a mine and sank. The German ship, a converted Hamburg-Amerika liner (2163 tons), had been sunk 105 km (60 miles) off Harwich on the previous day, but at 0630 the *Amphion* and her destroyers ran into the newly laid minefield. The mine blew away the entire forepart of the ship, and as she sank 20 minutes later a second mine detonated the forward magazine. Captain Fox was saved but 148 officers and men were killed.

The *Fearless* led the 1st Destroyer Flotilla of the Harwich Force, and fought at the Heligoland Bight action on August 28, 1914. In 1916 she became the leader of the 12th Submarine Flotilla ('K' Class steam-driven submarines) with the Grand Fleet until the Armistice. On January 31, 1918 she was badly damaged when she rammed and sank *K 17* during the notorious 'Battle of May Island' in the Firth of Forth, in a series of collisions during a night exercise with 'K' Class submarines. *Fearless* was disposed of in 1921 and broken up in Germany.

The *Active* led the 2nd Destroyer Flotilla at Harwich in 1914-15 and then at Scapa Flow. In 1916-17 she was leader of the 4th Destroyer Flotilla, on escort duties at Portsmouth, and then moved to Queenstown in 1917-18, and finally to the Mediterranean in 1918. She was sold in 1920 and broken up in Norway.

Displacement: 3440 tons (normal) *Length:* 123.4 m (405 ft 0 in) oa *Beam:* 12.5 m (41 ft 0 in) *Draught:* 4.3 m (14 ft 0 in) mean *Machinery:* 4-shaft steam turbines, 18 000 shp=25½ knots *Protection:* 25 mm (1 in) deck *Armament:* 10 4-in (100-mm) (10×1); 4 3-pdr (47-mm); 2 21-in (53-cm) torpedo tubes (underwater, broadside) *Crew:* 320

Activity

British escort carrier. The *Activity* was built as a merchant ship and taken over for conversion to a small aircraft carrier for service with convoys. In most respects she resembled other British CVEs (CVE is the US Navy designation for escort carriers) in being larger than the American conversions, and her twin-screw diesel engines enabled top speed to be maintained more easily.

The ship was ordered in 1941 from the Caledon Shipbuilding company of Dundee as the fast refrigerated cargo carrier *Empire Activity*. She was launched on May 30, 1942 as the *Telemachus*, completed on September 29 the same year, and was operated by the Alfred Holt Line. In 1943 she was requisitioned by the British Admiralty from the Ministry of War Transport for conversion to an escort carrier, but did not come into service until February 1944. She escorted convoys to Gibraltar and northern Russia during 1944 and 1945 and her Swordfish aircraft sank *U 288* on April 3, 1944.

The escort carrier *Activity*, a conversion of an unfinished merchant ship, shown taking on stores in April 1944. At that time she was engaged in escorting convoys to the Arctic carrying three Martlet (Grumman F4F Wildcat) fighters and nine Swordfish torpedo-bombers

The Air Group embarked in HMS *Activity*, 819 Squadron, operated three Swordfish and seven Wildcats in February 1944. This was increased to nine Swordfish and three Wildcats when the ship switched to Arctic convoys, as there was more need for antisubmarine aircraft than fighters. In 1946 she returned to mercantile service.

Displacement: 11 800 tons (standard), 14 300 tons (full load) *Length:* 156.27 m (512 ft 9 in) oa *Beam:* 20.26 m (66 ft 6 in) *Draught:* 7.92 m (26 ft) max *Machinery:* 2-shaft Burmeister & Main 6-cylinder diesels, 12 000 bhp=18 knots *Aircraft:* 15 max *Armament:* 2 4-in (102-mm) AA (1×2); 28 20-mm (14×2); originally only 10 single 20-mm carried *Crew:* 700

AD, AD-1 US Navy bomber See **Skyraider**

Adamastor

Portuguese cruiser. In 1895 a patriotic subscription was launched at Lisbon to buy a new ship for the Portuguese navy. Sufficient money was collected to order a small 3rd Class cruiser from the Italian Orlando shipyard, and the ship was launched in 1896 and completed the following year.

The *Adamastor* was a two-funnelled ship with a ram bow and counter stern. Her main armament was two German 10.5-cm guns, mounted fore and aft, with two 120-mm (4.7-in) guns in side sponsons a deck lower. Four 6-pdrs (57-mm), two 1-pdrs (25-mm) on the bridge and two Nordenfelt machine-guns in the fighting tops completed the gun armament and there was also an 18-in bow torpedo tube and two single swivelling tubes on the broadside above water. The hull was protected by a 1¼-in armoured deck and the conning tower by 2-in armour.

Two triple-expansion engines gave a speed of 17.19 knots on trials with 3000 ihp at natural draught. With forced draught she could make about one knot more.

In the 1920s she was given a complete refit, during which the original eight boilers were replaced by four, giving a speed of 18-19 knots with a forced draught. The armament was reduced to two 4.7-in, four 4.1-in (10.5-cm), four 3-pdr (47-mm) and three machine-guns. The torpedo armament remained unchanged, but the crew was reduced to 206. She was used on colonial police duties until sold in the mid-1930s.

Displacement: 1964 tons (normal) *Length:* 73.76 m (242 ft 3 in) pp *Beam:* 10.66 m (35 ft) *Draught:* 5.33 m (15 ft 3 in) *Armament, protection and machinery:* see above *Crew:* 235

Adder

US submarine class. *Adder* was the first of a class of seven US submarines launched between July 1901 and January 1903. They were improved versions of the prototype *Holland* (SS 1) which had been bought in April 1900.

The design was drawn up by the Electric Boat company of Groton, Connecticut, but the boats were built by the Crescent company *(Adder, Plunger, Moccasin, Porpoise* and *Shark)* and Union Iron Works *(Grampus* and *Pike)*. The *Adder* was the first to commission, on January 12, 1903, and the last two, *Porpoise* and *Shark*, commissioned in September 1903. All except the *Plunger* were transferred to the Philippines in 1909 as deck cargo aboard the colliers *Caesar* and *Hector*.

US Navy

Above: USS *Adder* undergoing sea trials in 1903. Developed from the original *Holland*, the *Adders* were the first class of submarines in the US Navy. *Below:* USS *Grampus*, another of the same class, in dry dock. In 1911 all submarines in US service were renamed with letters and numbers, *Adder* thus becoming *A 2* and *Grampus A 3*

US Navy

On November 17, 1911 all US submarines were given numbers instead of names, and as the *Holland* had been stricken the *Adder* Class became the A Class. In 1920 the US Navy adopted a standard hull-numbering system, and this was made retrospective to include even units that no longer existed. The A Class became SS 2-8. *A 5* was sunk at Cavite on April 15, 1917 after an explosion, and on July 24 of the same year *A 7* lost her entire crew in an engine explosion. The six surviving boats were stricken on January 16, 1922 (*A 1* had been stricken in February 1913) and all were sunk as targets.

Displacement: 107/123 tons (surfaced/submerged) *Length:* 19.42 m (63 ft 9 in) oa *Beam:* 3.65 m (12 ft) *Draught:* 3.19 m (10 ft 6 in) *Machinery:* (surfaced) single-shaft petrol engine, 160 bhp=8 knots; (submerged) electric motor, 150 ehp=7 knots *Armament:* 1 18-in (45-cm) torpedo tube (3 torpedoes) *Crew:* 7

Aden

British aircraft gun. In 1945 the Allies realized that German aircraft cannon were far more advanced than their own. In particular, the Mauser MG 213C 20-mm and MK 213C 30-mm guns promised performance unobtainable at that time by any other means. Like the US, USSR, France, Sweden and Switzerland, Britain copied the Mauser gun and produced the 30-mm Aden (Armaments Development ENfield). Prototypes were on test in 1947 and the Aden Mk 1 was ready for service in 1951—well ahead of the aircraft designed to carry it.

The first aircraft in service with the Aden was the Hunter F.1 in which four guns plus ammunition were mounted in a quickly replaceable pack hoisted into the fuselage by

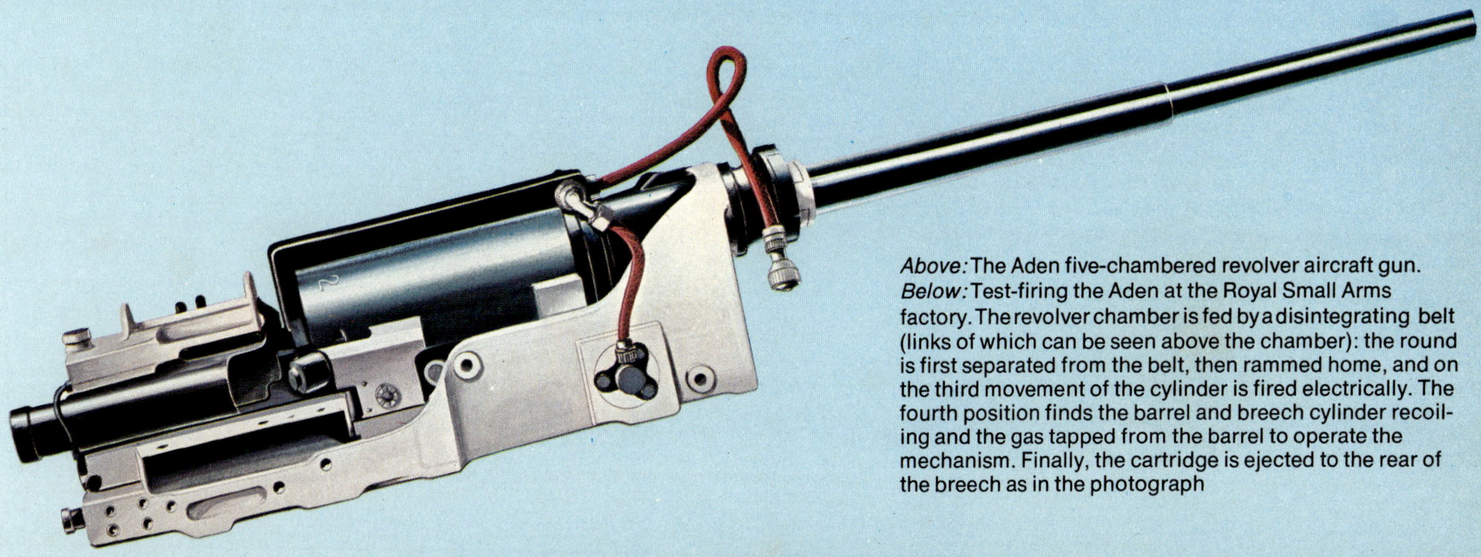

Above: The Aden five-chambered revolver aircraft gun. *Below:* Test-firing the Aden at the Royal Small Arms factory. The revolver chamber is fed by a disintegrating belt (links of which can be seen above the chamber): the round is first separated from the belt, then rammed home, and on the third movement of the cylinder is fired electrically. The fourth position finds the barrel and breech cylinder recoiling and the gas tapped from the barrel to operate the mechanism. Finally, the cartridge is ejected to the rear of the breech as in the photograph

winch. Other aircraft fitted with Aden 30-mm guns included the Swift, Gnat, Javelin, Scimitar, Lightning and Jaguar GR.1 and T.2.

The major production gun was the Mk IV. Like the French DEFA, this is still almost identical to its German ancestor. It is fed from either side by a metal link belt, has pneumatic cocking and charging, a five-chambered revolving cylinder and electrical firing of each round. Since 1960 the Aden Mk IV and an ammunition pack have been available as an external store for attachment to a weapon pylon. This is a standard fitment on the Harrier GR.1 and available for the Hawk.

Weight unloaded: 192 lb (87 kg) *Length:* 64.5 m (211 ft 7 in) *Operation:* Recoil and gas *Calibre:* 30 mm *Magazine:* Linked belt, variable length *Rate of fire:* 1200-1400 rpm *Muzzle velocity:* 2600 ft/sec (800 m/sec)

ADM-20 US decoy missile See **Quail**

RSAF Enfield

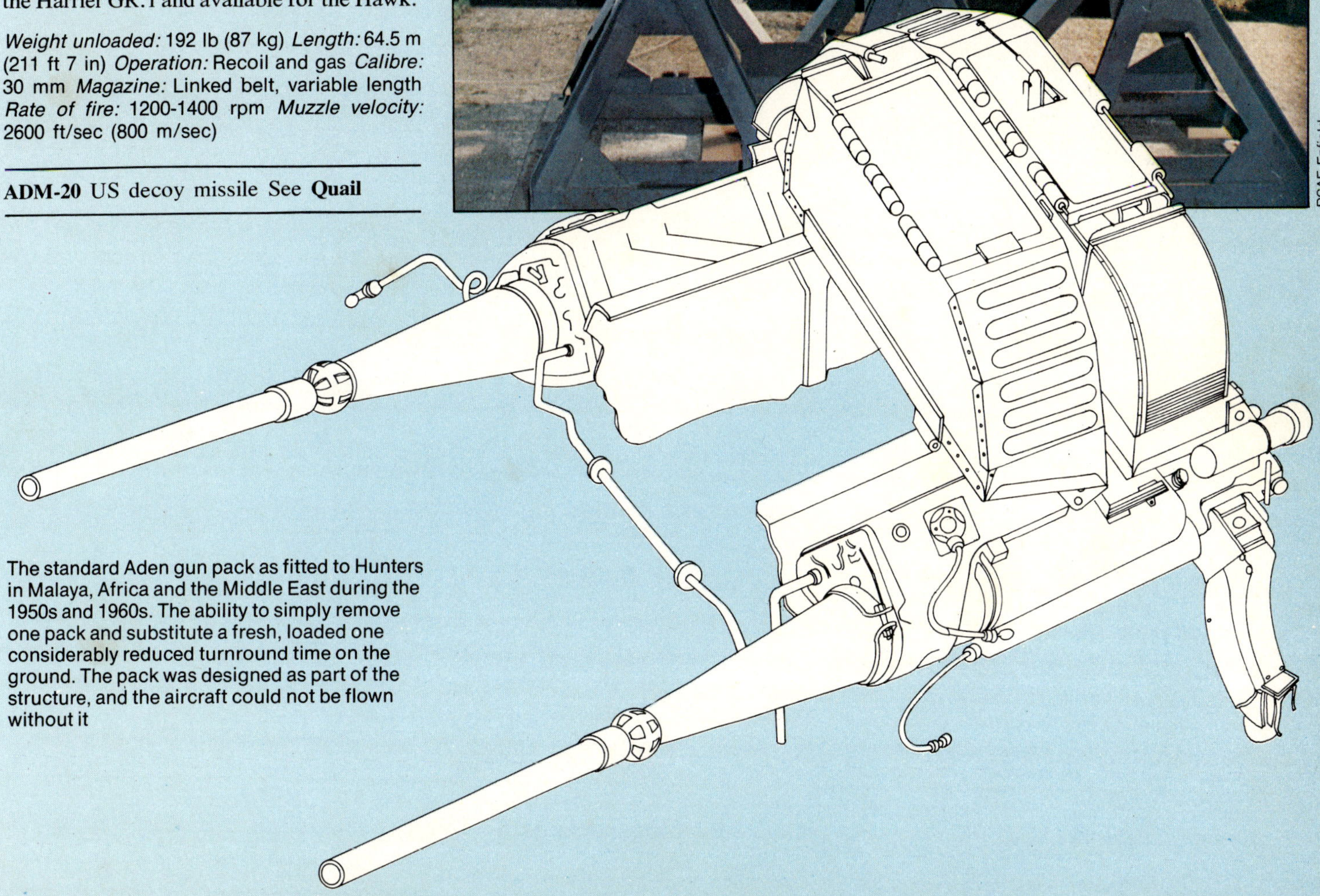

The standard Aden gun pack as fitted to Hunters in Malaya, Africa and the Middle East during the 1950s and 1960s. The ability to simply remove one pack and substitute a fresh, loaded one considerably reduced turnround time on the ground. The pack was designed as part of the structure, and the aircraft could not be flown without it

Admiral Butakov

Russian light cruiser class. Eight light cruisers were projected in 1912 for the Russian fleet as part of the massive programme of reconstruction to replace the losses of the war against Japan. They were built in two groups for the Baltic and Black Sea Fleets, three-funnelled ships with guns disposed fore and aft in broadside positions.

Admiral Greig and *Svetlana* were laid down at the Baltic works, Reval, in November 1913, *Admiral Butakov* and *Admiral Spiridov* at the Putilov works, St Petersburg, in December 1913, *Admiral Lazarev, Admiral Nachimov, Admiral Istomin* and *Admiral Kornilov* at Nikolaiev, in the Black Sea, between November 1913 and September 1914. Only four were launched by the time of the Revolution, and of these only three were finally completed, to much altered designs. *Admiral Lazarev* became *Krasni Kavkaz, Admiral Nachimov* became *Chervonaya Ukraina* and *Svetlana* became *Krasni Krim.*

The remaining ships suffered a variety of vicissitudes. *Admiral Butakov* was stopped in 1916-17 but her hull survived the Second World War and was scuttled as a breakwater at the mouth of the River Neva. She was scrapped in 1952.

Admiral Spiridov lay on the stocks until 1926, when she was completed as the mercantile tanker *Groznyeft.* Renamed *Grozny* in 1935, she was captured by the Germans in October 1941 and was scuttled at Mariupol in 1943.

Admiral Istomin was abandoned on the slip in 1917, but not broken up until many years later.

Admiral Greig was completed as the mercantile tanker *Azneft* in 1926. She foundered in a storm in the Black Sea in the 1930s.

Admiral Kornilov was only about 40% complete in 1917, and was scrapped about 1930.

See also: *Krasni Kavkaz, Chervonaya Ukraina* and *Krasni Krim.*

Displacement: 6800 tons (normal) *Length:* 157.20 m (519 ft 9 in) oa *Beam:* 15.34 m (50 ft 4 in) *Draught:* 5.59 m (18 ft 4 in) *Machinery:* 2-shaft steam turbines, 50 000 shp=$29\frac{1}{2}$ knots *Protection:* 75 mm (3 in) belt, 25 mm (1 in) deck, 75 mm (3 in) gunshields *Armament:* 15 5.1-in (130-mm) guns (15×1); 4 4-in (100-mm) AA (4×1); 4 3-in (75-mm) (4×1); 2 submerged torpedo tubes (probably 21-in) *Crew:* 600 approx

Admiral Graf Spee

German pocket battleship. The *Admiral Graf Spee* (usually known as the *Graf Spee*) and her sister *Admiral Scheer* were beamier versions of the prototype *Deutschland,* and had less range. Popularly known as 'pocket battleships', they were actually known to the German navy as *Panzerschiffe,* or armoured ships, and were in fact only very heavily gunned heavy cruisers. They were an answer to the restrictions of the Versailles Treaty which followed Germany's defeat in 1918. Germany was not allowed to build warships displacing more than 10 000 tons or armed with guns larger than 280 mm (11 in), with the avowed intention of limiting her to coast defence ships.

But the German constructors, by using

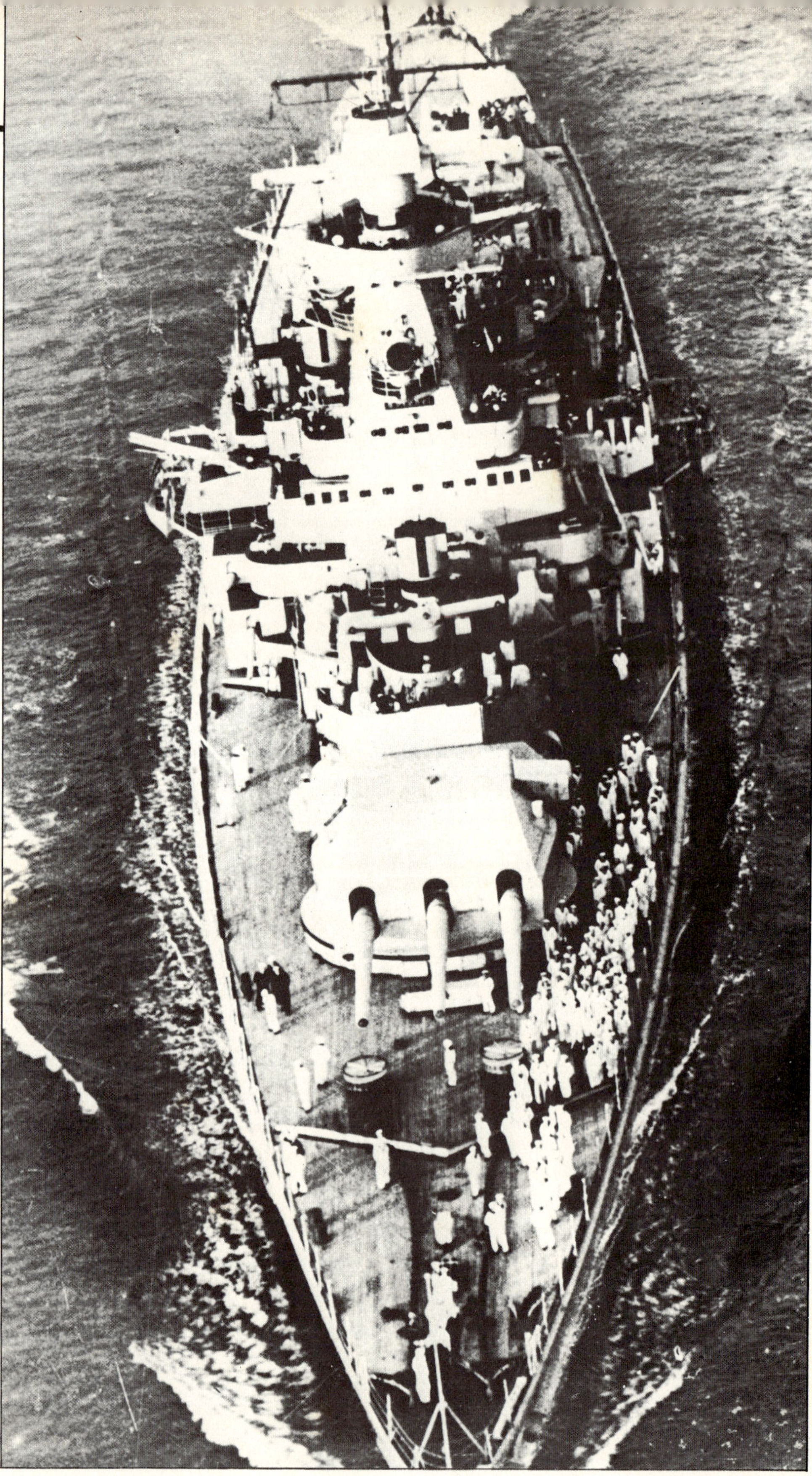

The *Admiral Graf Spee* was the third and last of Germany's *Panzerschiffe* (armoured ships), or 'pocket battleships'. Electrically welded to save weight and given diesel engines for speed, they were designed as commerce raiders fast enough to escape any ship they could not outgun. *Graf Spee* is seen (above) from the air and (below) on trials in 1936

IWM

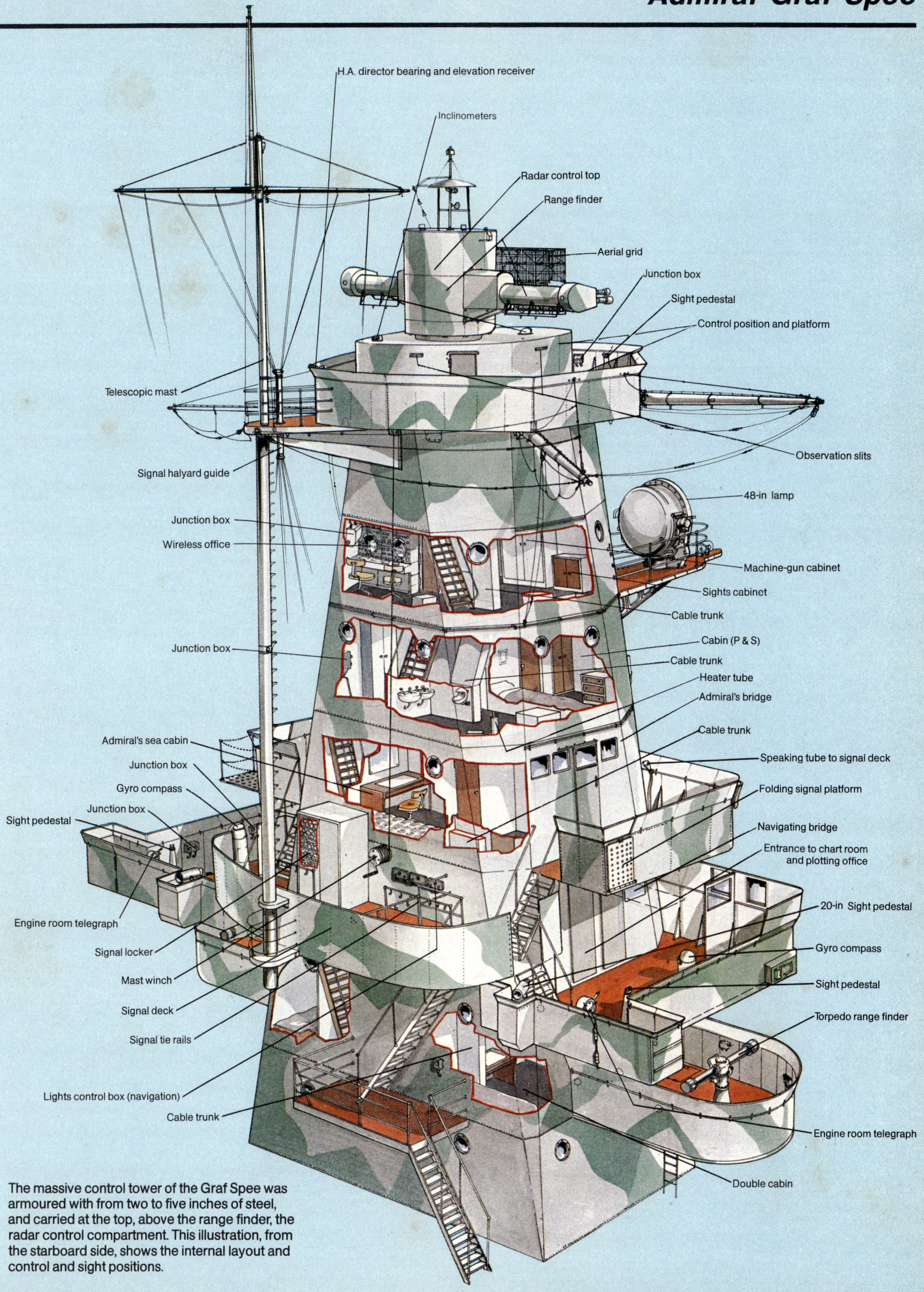

The massive control tower of the Graf Spee was armoured with from two to five inches of steel, and carried at the top, above the range finder, the radar control compartment. This illustration, from the starboard side, shows the internal layout and control and sight positions.

lightweight construction and welding, as well as economical diesel engines, but above all by concealing the true tonnage, produced a warship with heavy gunpower and impressive range. This caused a sensation among the world's navies. Here, apparently, was the ideal commerce-raider, faster than anything more powerful (apart from a handful of battlecruisers) and more powerful than anything faster.

However, no other navy copied the design and the Germans built only three of them. The diesels were not quite as successful as hoped, and the new generation of fast battleships with speeds of 28-30 knots threatened the rationale of the design. Nor were their fighting qualities all that was required. The concentration of the heavy armament in two turrets was a weak point, while the protection was little better than that of the 8-in (200-mm) gunned 'treaty cruisers' that were their most likely opponents. In short, the pocket battleship was a needlessly large and expensive commerce-raider, for whom the most important rule was to avoid action with any warship which could fight back.

The *Graf Spee* was the last of the type to be built, being ordered under the 1932 Programme and launched in 1934. She differed from the *Scheer* in having a thinner but deeper armour belt, and in carrying slightly more fuel. She was at sea when war broke out in September 1939, having already proceeded to her war station in the Atlantic. During a moderately successful cruise in the South Atlantic she sank nine ships (50 089 tons), without a single life being lost. However, early on the morning of December 13, 1939, she sighted three British cruisers off Montevideo, the 8-in (200-mm) gunned *Exeter* and the 6-in (152-mm) gunned *Ajax* and *Achilles*.

Captain Langsdorff decided to engage. He had the advantage of a primitive radar set, but all British cruisers had practised antiradar tactics for years past and, in accordance with well-established doctrine, the *Exeter* and her two consorts split up to divide the German fire. During the 'Battle of the River Plate' which followed, the *Exeter* was badly damaged, while the *Ajax* was not in much better state, but the *Graf Spee* fired away most of her ammunition and suffered sufficient damage to make her put into Montevideo for repairs. Once there, her captain was bluffed into thinking that the British had gathered reinforcements, and four days later she steamed out to be scuttled by charges detonated in the main magazines and the engine room.

See also *Deutschland* and *Admiral Scheer*.

Displacement: 12 000 tons (standard), 16 200 tons (full load) *Length:* 187.91 m (616 ft 9 in) oa *Beam:* 21.72 m (71 ft 3 in) *Draught:* 7.32 m (24 ft 0 in) max *Machinery:* 2-shaft diesels (8 MAN), 57 800 bhp=26 knots *Protection:* 57-82 mm (2¼ in-3¾ in) belt, 19-76 mm (¾ in-3 in) decks, 13 mm-130 mm (½ in-5 in) turrets, external antitorpedo bulge *Armament:* 6 11-in (280-mm) (2×3); 8 5.9-in (150-mm) (8×1); 6 4.1-in (105-mm) (3×2); 8 37-mm (1.5-in) AA (4×2); 10 20-mm (0.79-in) AA (10×1); 8 21-in (53-cm) torpedo tubes (2×4); 2 aircraft (1 catapult) *Crew:* 1124

The *Graf Spee* in the South Atlantic in 1939: a false turret has been fitted as a disguise. The Spee coat of arms (inset) was carried on her bow but in wartime it was painted over

Admiral Hipper

German heavy cruiser. The *Admiral Hipper* and her sister *Blücher* were ordered under the 1934 Programme, the first heavy cruisers built for the German navy since before 1914. They served as the model for the further class of three, of which *Prinz Eugen* was the only one to be completed (one was handed over to the Russians).

Although theoretically 10 000-ton ships, they displaced nearly twice as much as some of the earlier 'treaty cruisers' of other navies. They were bigger, and in some ways more powerful, than the 'pocket battleships'. Although intended for raiding, their endurance was too low for this task, even if their high-pressure steam machinery had given less trouble than it did. The armour was not well distributed but the antiaircraft armament was good, especially after the *Admiral Hipper*, which was built with two antiaircraft

The *Admiral Graf Spee* in her final form in the grey and olive-green South Atlantic camouflage scheme. The false bow waves along the waterline were intended to confuse submarines by making her speed impossible to calculate

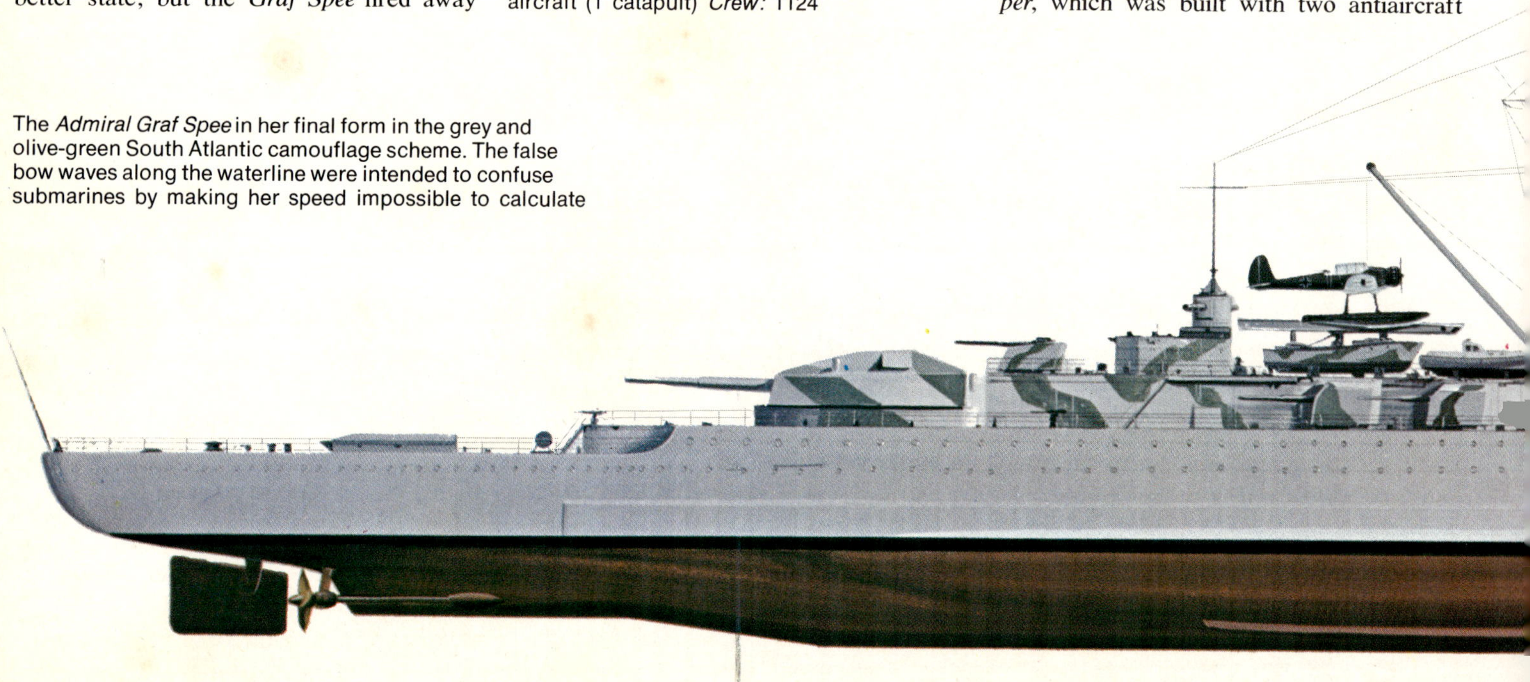

directors, was given an extra pair early in the war, and an additional 35 20-mm (0.79-in) guns. Despite her mechanical weaknesses, she was a formidable fighting ship.

The appearance of the ship was altered during the same refit. She had been completed with an extremely ugly flat-topped funnel and a straight bow but at the outbreak of war was rebuilt with a raked and lengthened 'Atlantic' bow and capped funnel, making her look like most of the other German heavy ships.

The *Admiral Hipper* was launched at the Hamburg yard of Blohm & Voss in February 1937. After an unsuccessful sortie early in 1940 her first action was during the Norwegian campaign when she led a group of destroyers carrying troops to Trondheim. On the way she was attacked by the British destroyer *Glowworm*, which managed to ram her forward, causing considerable damage before being sunk. Later in the same campaign the *Hipper*'s low endurance and machinery troubles forced her to part company with the battlecruisers *Scharnhorst* and *Gneisenau* before they caught and sank the British aircraft carrier *Glorious*.

In December 1940 she went on a raiding cruise in the Atlantic, emerging successfully through the Denmark Strait. Her orders were to attack convoys but the first convoy which she sighted was a troop convoy escorted by cruisers and the aircraft carrier HMS *Furious*. Not realizing this, she attacked at dawn but retired after receiving slight damage, and made off before the escort could bring her to action. She then became the first German heavy unit to use the French Atlantic port of Brest.

In February 1941 she left for another cruise, and this time she was lucky enough to catch an unescorted group of 19 ships, and sank seven of them, totalling 32 806 tons. But she was low on fuel and had to return to Germany after calling at Brest. In 1942 she was sent to Norway and on December 31, in company with the *Lutzow* and six destroyers, took part in an attack on the British convoy JW51B bound for north Russia. The ensuing 'Battle of the Barents Sea' virtually finished the German navy in Hitler's eyes, for the escort of destroyers kept the cruiser and the pocket battleship at bay until two British 6-in (152-mm) gunned cruisers came up and chased them away.

The *Admiral Hipper* was damaged near the end of the war by RAF bombs, and was scuttled at Kiel on April 3, 1945 to avoid capture. The hull was eventually broken up in 1948.

PPL

The German heavy cruiser *Admiral Hipper* (above) in her original form with straight bow and flat-topped funnel. Early in the war she was rammed by the destroyer *Glowworm*, and during the repairs her bow was raked and the funnel capped. *Below:* In harbour later in the war

IWM

Displacement: 10 000 tons (nominal standard), 12 500 tons (actual standard), 13 900 tons (normal), 17 250 tons (full load) *Length:* 206 m (675 ft 9 in) oa *Beam:* 21 m (70 ft) *Draught:* 8.4 m (25 ft 3 in) max *Machinery:* 3-shaft geared steam turbines, 132 000 shp=32.5 knots *Protection:* 60 mm-80 mm ($2\frac{1}{4}$ in-$3\frac{3}{4}$ in) belt; 30 mm-50 mm ($1\frac{1}{4}$ in-2 in) decks; 70 mm-110 mm ($2\frac{3}{4}$ in-$4\frac{1}{4}$ in) turrets; 50 mm (2 in) conning tower *Armament:* 8 8-in (200-mm) (4×2); 12 4.1-in (105-mm) (6×2); 12 37-mm (1.5-in) AA (6×2); 4 20-mm (0.79-in) AA guns; 12 21-in (53-cm) torpedo tubes (4×3) *Aircraft:* 3 (1 catapult) *Crew:* 1600

A broadside from the *Admiral Scheer*'s 11-in guns. The *Scheer* was the second of Germany's three *Panzerschiffe* or 'pocket battleships'

Admiral Makarov

Russian armoured cruiser class. Three large armoured (ie protected by side armour) cruisers were built in 1905-10 to a French design to replace the losses of the Russo-Japanese war. In fact they were very similar to the *Bayan*, which fell into Japanese hands at Port Arthur.

As the Baltic Fleet was short of battleships early in the First World War, the big cruisers were used as the spearhead of the Fleet. *Admiral Makarov* and the older *Gromoboi* drove off a German minelaying force on August 17, 1914 and nine days later the *Pallada* helped to drive the cruiser *Magdeburg* ashore in the Eastern Baltic. It was this action which caused the German codes and cyphers to fall into Allied hands.

On October 10 the submarine *U 26* fired two torpedoes at the *Admiral Makarov* without success, but the next day she hit the *Pallada*, which blew up and sank with all hands. In mid-December the *Admiral Makarov* and *Bayan* laid mines off Danzig, one of the last operations before the Baltic froze over. In May 1915 the two cruisers covered another minelaying operation. Early on the morning of July 2 the *Admiral Makarov* (flagship of Rear-Admiral Bakhirov) and the *Bayan* were in action against the German armoured cruiser *Roon* and the light cruiser *Lübeck*. The *Bayan* was hit by the *Roon* but managed to inflict slight damage on the German ship in return before contact was lost. The Baltic Fleet was in action later in 1915 but its efficiency was being steadily reduced, and the major warships did little in 1916. The two cruisers were among the many ships which joined the Revolution in 1917, and when the time came to rebuild the Red Fleet after the upheavals of the civil war they were found to be in very bad condition. Both ships were sold to German shipbreakers for scrapping and they were finally broken up at Stettin in 1922.

Admiral Makarov launched May 1906 at La Seyne yard, Toulon
Pallada launched November 1906 at Galernii Island, St Petersburg
Bayan launched August 1907 by New Admiralty Works, St Petersburg

Displacement: 7835 tons (normal) *Length:* 136.8 m (449 ft 9 in) oa *Beam:* 17.3 m (57 ft) *Draught:* 6.5 m (21 ft 3 in) *Machinery:* 2-shaft triple-expansion, 16 500 ihp=21 knots *Protection:* 100 mm-180 mm (4-7 in) belt; 50 mm (2 in) deck; 180 mm (7 in) turrets *Armament:* 2 8-in (200-mm) (2×1); 8 6-in (152-mm) (8×1); 22 75-mm (3-in); 4 47-mm (1.8-in); 2 75-mm and 2 47-mm AA guns added 1915-16 and mine rails for 150 mines, except in *Pallada* *Crew:* 568

Admiral Scheer

German pocket battleship. The *Admiral Scheer* was very similar to the *Admiral Graf Spee*, apart from having a slightly thicker armour belt (100 mm) which was shallower, and slightly less fuel. She was ordered as part of the 1931 Programme, and was the second of three *Panzerschiffe* built by the German navy. She was slightly beamier than the prototype *Deutschland*, and, like the *Graf Spee*, differed in having a more conspicuous tower bridge and an aircraft catapult abaft the funnel. Unlike the *Graf Spee*, she lasted long enough to be refitted with a raked 'Atlantic' bow to improve her seaworthiness, and a funnel cap which made her look rather more rakish. The original 20-mm (0.79-in) guns were increased to 36-mm (1.4-in) during the war but were later replaced by 6 twin 37-mm (1.5-in) guns.

The ship was refitting during the Norwegian campaign and so it was not until October 1940 that she sailed on her first war mission into the Atlantic. She caught the convoy HX84 which was escorted by a single armed merchant cruiser, *Jervis Bay*, but due to the heroic self-sacrifice of that ship, the *Scheer* was only able to sink five of the convoy.

By January 1941 the *Admiral Scheer* was operating in the Indian Ocean, where she had some success by pretending to be a British warship. Later she sailed to the Atlantic and then to Germany. When she returned on April 1, 1941 she had steamed 46 419 miles and had sunk 16 ships, totalling 99 059 tons (including the *Jervis Bay*). In 1942 she was moved to Norway, where her presence posed a threat to British convoys bound for north Russia. However, she only took part in a couple of abortive operations, and was relegated to the training squadron in the Baltic in 1943. In 1944 she fired at advancing Russian troops, and was sunk on April 9, 1945 by RAF bombs at Kiel. The hull capsized, and proved so badly damaged that it was eventually buried under thousands of tons of rubble when the basin was filled in after the war.

Displacement: 12 100 tons (standard) *Length:* 185.6 m (609 ft 0 in) oa *Beam:* 21.3 m (70 ft 0 in) *Draught:* 6.5 m (21 ft 6 in) *Machinery:* 2-shaft diesels, 56 000 bhp=26 knots *Protection:* 100 mm (4 in) belt, 50 mm-100 mm (2 in-$5\frac{1}{2}$ in) turrets, 40 mm-75 mm ($1\frac{1}{2}$ in-3 in) decks, 50 mm-130 mm (2 in-5 in) conning tower *Armament:* 6 11-in (280-mm) (2×3); 8 150-mm (5.9-in) (8×1); 6 105-mm (4.1-in) AA (3×2); 8 37-mm (1.5-in) AA (4×2); 10 20-mm (0.79-in) AA; 8 21-in (53-cm) torpedo tubes (2×4) *Crew:* 1150

Admiral Spaun

Austro-Hungarian scout cruiser. This scout or light cruiser resembled contemporary British cruisers, both in appearance and characteristics. She was laid down at the Arsenal, Pola (now Pula in Yugoslavia) in May 1908 and commissioned in November 1910. In August 1914 she was the leader of the 2nd Torpedo Flotilla, comprising six destroyers and 18 torpedo boats.

In 1917 the *Admiral Spaun* was armed with a 66-mm (2.5-in) L50 antiaircraft gun. She and the *Saida* Class were outranged by British cruisers' 6-in (152-mm) guns in the Adriatic, and so in 1918 it was proposed that she should be rearmed with two 150-mm (5.9-in) guns and six 100-mm (3.9-in). No further details are known, and presumably the war ended before the work could be done. She was allocated to Britain after the Armistice, and was subsequently sold to Italian shipbreakers in 1920.

Displacement: 3500 tons (normal), 4007 tons (full load) *Length:* 125.2 m (410 ft 9 in) *Beam:* 12.8 m (42 ft 0 in) *Draught:* 5.3 m (17 ft 5 in) max *Machinery:* 4-shaft Parsons steam turbines, 25 250 shp=27 knots *Protection:* 60 mm (2.4 in) side, 20 mm (0.79 in) deck *Armament:* 7 100-mm (3.9-in) guns; 2 machine-guns; 8 18-in (46-cm) torpedo tubes *Crew:* 320

Adua

Italian submarine class. The *Adua* Class of 600-ton submarines were designed by General Curio Bernardis, and were developed from the earlier *Argonauta, Sirena* and *Perla* Classes. Seventeen units were laid down in 1936-37 and completed in 1936-38, in addition to another three built for Brazil in 1937.

In 1940-41 *Gondar* and *Sciré* were fitted as transports for 'human torpedoes' or *Maiale* (pigs), with the 100-mm gun removed and three cylinders welded onto the deck-casing. All but one of the class became casualties during the Second World War.

JA Roberts

HMS *Adventure*, the Royal Navy's first purpose-built minelayer, was a somewhat unsuccessful design. Too lightly armed for a genuine cruiser, she was not fast enough for offensive minelaying operations. She is shown (above) in May 1932 and (below) in February 1943

MOD

Displacement: 680-698/848-866 tons (surfaced/submerged) *Length:* 60.18 m (197 ft 5 in) oa *Beam:* 6.45 m (21 ft 2 in) *Draught:* 4.69-4.75 m (15 ft 6 in-9 in) *Machinery:* (surfaced) 2-shaft diesels, 1200 bhp=14 knots; (submerged) electric motors, 800 ehp=$7\frac{1}{2}$ knots *Armament:* 1 100-mm (3.9-in) gun, 2-4 13.2-mm (0.51-in) guns, 6 21-in (53-cm) torpedo tubes (12 torpedoes) *Crew:* 46

Adventure

British minelayer. The *Adventure* was the first purpose-built minelayer constructed for the Royal Navy. She was laid down in November 1922 at Devonport Dockyard, launched in June 1924 and commissioned in May 1927.

She was intended for offensive operations in which she could use her speed to enter enemy waters and lay a minefield without being intercepted. However, her speed was insufficient for this purpose, being well below that of most cruisers. She was, because of her size, designated as a cruiser/minelayer, but her armament hardly placed her in the cruiser class. The most interesting feature of her design was the installation of a diesel-electric drive (manufactured by Vickers-Armstrong) for cruising speeds, which were extremely economical and gave her a substantial radius of action.

HMS *Adventure* served in the Atlantic and Home Fleets until 1931 when she was taken in hand at Devonport for a refit. She was recommissioned in 1933 for the China station where she served for three years. After temporary service in the Mediterranean in 1936, she returned home and was placed in reserve until 1939, when she rejoined the Home Fleet.

Her first action was to assist in laying a defensive minefield in the Straits of Dover during September and October 1939. Ironically, in November, she was damaged by a magnetic mine and was out of commission for some months while being repaired. On April 10, 1943, while returning home from the Mediterranean, she encountered the German blockade runner *Irene* off Finisterre, and the German ship promptly scuttled herself to avoid being captured.

Besides laying mines the *Adventure* was used to carry supplies; in July 1941 she sailed to Archangel with a cargo of mines for the Russians and at the end of 1942 she twice carried 2000 aircraft depth-charges to Gibraltar. (The depth-charges were eventually taken to Malta where they were urgently needed.) In 1944 the *Adventure* was converted to a landing craft repair ship and operated in that role during the invasion of Normandy. On May 14, 1945 she was the first Allied ship to enter Cuxhaven harbour. At the end of 1945 she was placed in reserve and was eventually sold for scrapping in 1947.

Displacement: 6740 tons *Length:* 158.5 m (520 ft) *Beam:* 17.98 m (59 ft) *Draught:* 4.42 m (14 ft 6 in) *Machinery:* 4-shaft geared turbines (diesel electric for cruising), 40 000 shp=27¾ knots *Protection:* 50 mm (2 in) side; 25 mm (1 in) deck over machinery *Armament:* 4×4.7-in AA; 4 2-pdr pom-poms (added 1939); 340 (small) or 280 (large) mines *Crew:* 395

Adventure

British scout cruiser class. At the turn of the century the Royal Navy designed a series of small cruisers suitable for scouting in the North Sea. Armoured cruisers and even 2nd Class cruisers were too large, and existing 3rd Class cruisers were too slow for scouting and working with destroyers. As the new cruisers were considerably faster than existing types, the leading builders were invited to meet a broad Admiralty specification with their own designs.

The *Adventure* and *Attentive* were the pair ordered from Armstrongs' Elswick yard on the Tyne, and they differed from the *Foresight*, *Pathfinder* and *Sentinel* groups in having four funnels instead of three. The *Adventure* was originally to be named *Eddystone*; she and her sister were laid down in January 1904 and completed in October 1905. In 1911-12 they and their six sisters were rearmed with 9 4-in (102-mm) guns in place of the 14 12-pdr (3-in) guns, to cope with the growing size and armament of destroyers.

Both ships spent much of their time with destroyers until relieved by flotilla leaders in 1915. *Adventure* was the flagship of Admiral Sir Lewis Bayly at Queenstown in 1915-17. *Attentive* served with the Dover Patrol and then went to the White Sea in 1918.

Displacement: 2640 tons (normal), 3210 tons (full load) *Length:* 120 m (395 ft) oa *Beam:* 11.5 m (38 ft 3 in) *Draught:* 7 m (23 ft 0 in) max *Machinery:* 2-shaft vertical triple-expansion, 16 000 ihp=25 knots *Protection:* 20 mm-50 mm (¾ in-2 in) deck; 75 mm (3 in) conning tower *Armament:* (Designed) 10 12-pdr (10×1); 8 3-pdr (8×1); 2 18-in (46-cm) torpedo tubes; (Shortly after completion) 14 12-pdr; 6 6-pdr; 2 18-in torpedo tubes; (As altered 1911-12) 9 4-in (102-mm); 6 6-pdr; (Wartime) Both ships had a 3-in (76-mm) antiaircraft gun added, and in 1918 *Attentive* had 2 6-in (152-mm), 6 4-in etc *Crew:* 268

US Navy

The Japanese armoured cruiser *Adzuma* in the early 1900s before she was damaged

Adzuma

Japanese cruiser. This armoured cruiser was built in St Nazaire by Ateliers et Chantiers de la Loire between March 1898 and July 1900. Although her hull, machinery and protection were French-designed, the Japanese navy stipulated that the armament should be supplied from England.

In February 1904 the *Adzuma* bombarded Port Arthur and later blockaded Vladivostok. On August 14, 1904 she fought in the Battle

of Ulsan against the Russians, during which she was hit ten times and suffered eight casualties. In January 1905 she and the *Asama* blockaded the Straits of Tsugaru, and on May 27, 1905 she fought in the battle line at Tsushima. She was hit 16 times by Russian shells and suffered 11 dead and 29 wounded.

From 1914 she was relegated to training, and in September 1921 she was rerated as a 1st Class coast defence ship, with reduced armament. In 1941 she was hulked as a training ship at Yokosuka, and was badly damaged by American air attack on July 18, 1945. She was surrendered in a badly damaged state and was scrapped in 1946.

Displacement: 9307 tons (normal) 9943 tons (full load) *Length:* 137.8 m (452 ft 6 in) oa *Beam:* 17.7 m (68 ft 9 in) *Draught:* 7.5 m (24 ft 9 in) max *Machinery:* 2-shaft vertical triple-expansion, 17 000 ihp=20 knots *Protection:* 90 mm-180 mm ($3\frac{1}{2}$ in-7 in) belt; 25 mm-60 mm (1 in-$2\frac{1}{2}$ in) deck; 250 mm (14 in) conning tower *Armament:* 4 8-in (203-mm) (2×2); 12 6-in (152-mm) (12×1); 12 12-pdr (12×1); 12 3-pdr (12×1); 5 18-in (46-cm) torpedo tubes *Crew:* 650

AEC

British armoured cars. The Associated Equipment Company of Southall, Middlesex, were producers of heavy truck and bus chassis, and were well known during the Second World War for their 'Matador' gun tractor, which was the standard prime mover for medium field and heavy antiaircraft artillery. In 1941-42, using 'Matador' components as a basis, they developed the AEC Mk 1 armoured car, an 11-ton vehicle carrying a Valentine tank turret which mounted a 2-pdr gun and a coaxial 7.92-mm (120-mm) Besa machine-gun. These went into action in the latter part of the North African campaign with considerable success.

Encouraged by this, the company then developed their Mark 2 version, which used a larger turret and mounted a 6-pdr gun, and

RAC Tank Museum

The AEC Mk I was designed to have equivalent armour and armament to a tank: appropriately, the 158-bhp engine of the Mk II (above) was like that of a Valentine tank

RAC Tank Museum

The AEC Mk III (above) was outwardly similar to the Mk II, but substituted a 75-mm gun for the latter's 6-pdr. The 'Armoured Car, AEC, Mark I' (left) a private venture by the Associated Equipment Company, was the first British production car to use a diesel engine (7.58 litre, delivering 105 bhp). A conspicuously-painted mock-up was unofficially introduced into a display of vehicles at Horse Guards Parade in 1941, caught the prime minister's attention and won a production order

RAC Tank Museum

RAC Tank Museum

The final version of the AEC was this prototype antiaircraft version, mounting two 20-mm Oerlikon cannon in a large turret. Developed in 1944, it never achieved production status

then the Mark 3 'Close Support Armoured Car', which carried a medium-velocity 75-mm (3-in) gun. The Mark 3 was larger and heavier (12.7 tons), and was less successful since it was underpowered and had a poor cross-country performance. Nevertheless, it remained in service until replaced by the Saladin armoured car in the 1950s.

(Mk 1) *Weight:* 11 tons *Length:* 5.18 m (17 ft) *Width:* 2.74 m (9 ft) *Height:* 2.54 m (8 ft $4\frac{1}{2}$ in) *Speed:* 58 km/h (36 mph) *Crew:* 3 *Range:* 402 km (250 miles)

(Mk 2) *Weight:* 12.7 tons *Length:* 5.43 m (17 ft 10 in) *Width:* 2.69 m (8 ft $10\frac{1}{2}$ in) *Height:* 2.68 m (8 ft 10 in) *Speed:* 66 km/h (41 mph) *Crew:* 4 *Range:* 402 km (250 miles)

(Mk 3) *Weight:* 12.7 tons *Length:* 5.60 m (18 ft 5 in) *Width, height, speed, crew* and *range* as for Mark 2

Introduced in 1915, the AEG C.IV was a two-place single-engined scout, armed with a Parabellum machine-gun in the rear cockpit and a forward-firing synchronized spandau

The most numerous AEG type was the G.IV, introduced in the second half of 1916. Having only a 350-kg (770-lb) bombload, it was largely restricted to short-range tactical bombing or photographic reconnaissance missions. The octagonal camouflage scheme shown employing printed fabric was used on German bombers throughout the war

AEG

German aircraft series. One of the largest German electrical engineering companies, the Allgemeine Elektrizitäts Gesellschaft of Berlin-Henningsdorf, was a major supplier of large warplanes in the First World War. The first AEG, the K.I (K for *Kampfflugzeug* or combat aircraft) of February 1915, was a three-seater biplane with two 100-hp engines. The G.II was slightly bigger and had 150-hp Benz Bz III engines and could carry a bomb load of up to 200 kg (440 lb). A clean three-bay biplane, the G.II had either a single large fin and rudder or a triple-finned tail, both versions serving together in the *Schlachtstaffeln* (Battle Flights) on the Western Front.

In the winter of 1915-16 the still larger G.III appeared, with 220-hp Mercedes engines and a 300-kg (660-lb) bomb load. This was supplanted in the autumn of 1916 by the G.IV, roughly the same size (data below), but heavier, and with 260-hp Mercedes D.IVa engines which were not prone to the breakdowns of the geared 220-hp engines. The G.IV was delivered in several versions, and they served as bombers over ranges of up to

Armed with two Parabellum machine-guns, the AEG G.III had a 300-kg (660-lb) bombload

650 kilometres (400 miles)—too short to bomb London—and as reconnaissance aircraft. The G.IVk had the 20-mm (0.79-in) Becker cannon and biplane tail, and the G.IVb had a greater span.

The final AEG model was the G.V, with the same 260-hp engines but much larger span and a bomb load of 600 kg (1320 lb). Only a few of these were built before the Armistice, and it is believed that as many as 400 of the total AEG production of 542 bombers were aircraft of the basic G.IV type.

Most were painted in the multi-coloured lozenge camouflage schemes used by the Imperial Military Aviation Service, but a few were finished in irregular purple/violet/black camouflage or in other schemes, including white overall.

Span: 18.3 m (60 ft 2 in) *Length:* 9.9 m (32 ft 4 in) *Gross weight:* 3630 kg (8003 lb) *Max speed:* 145 km/h (90 mph) incorrectly reported as 166 km/h

Aermacchi Italian aircraft See **MB 326, 339**

Aeritalia Italian aircraft See **G91, G222**

Aerospatiale French aircraft See **Alouette, Gazelle, Lynx, Magister, Puma, Super Frelon**

Aero Czech aircraft See **A11, Albatross, Delfin, MB 200**

Aetos

Greek destroyer class. As part of her preparations for war against Turkey in the years before the First World War, Greece ordered a number of destroyers from French, German and British shipyards. Four had been ordered for Argentina from the British firm of Cammell Laird, and were launched in 1911, but the Greek government bought them in October 1912, renaming them *Aetos, Leon, Ierax* and *Panthir* (ex-*San Luis, Santa Fe, Tucuman* and *Santiago*).

They were handsome vessels with five funnels, two paired forward, a single, and two paired aft. They resembled contemporary British destroyers but were armed with four American-pattern 4-in (102-mm) guns supplied by Bethlehem Steel.

By the end of the First World War they were obsolescent, and in 1924-25 the British firm of J Samuel White modernized all four destroyers. The reconstruction was complete, changing them into two-funnelled craft resembling the latest British destroyers of the Modified 'W' type. After the refit they exceeded their original trials speeds by 2 knots, and they rate as one of the most successful attempts at modernization.

The *Leon* was sunk in action with German aircraft on April 22, 1941 shortly after the invasion of Greece, but the remaining three ships survived the Second World War and were later scrapped.

Displacement: (As built) 980 tons (normal) (As modernized) 1013 tons (normal), 1300 tons (full load) *Length:* 89.3 m (293 ft) *Beam:* 8.4 m (27 ft 9 in) *Draught:* 3 m (10 ft) full load *Machinery:* 2-shaft Parsons turbines, 19750 shp=32 knots *Armament:* (As built) 4 4-in (102-mm) (4×1); 4 21-in (530-mm) torpedo tubes (2×2); (As modernized) 4 4-in (4×1); 2 2-pdr (2×1); 6 21-in (53-cm) torpedo tubes (2×3) *Crew:* 110

Afridi

British destroyer class. During the 1930s several countries built large, fast destroyers with a heavy gun armament. The *Afridi* Class, which were more commonly known as the 'Tribals', were built to counter these foreign ships. Compared with earlier British destroyers they carried twice as many guns and were almost 500 tons heavier. It was, however, necessary to restrict their torpedo armament to four 21-in tubes (earlier ships carried eight or ten tubes) in order to reduce cost and keep them within the international treaty limit of 1850 tons. This limitation lapsed before the ships were completed and additions during construction increased their displacements by over 100 tons. They were the first British destroyer class to carry twin gun mountings and the last to be constructed on the transverse framing system. The 16 Royal Navy ships were completed between May 1938 and March 1939. The four Royal Canadian Navy vessels built in Britain were ordered in 1940-41 and completed in 1942-43. The four built in Canada were ordered in 1942

PPL

Eskimo in February 1939. The 'F' pendant numbers became 'G's in 1940

and laid down in the following year, but did not complete until after the war. Australia originally intended to build seven but in the event only laid down three. The first pair completed in 1942 and served in the Pacific, but *Bataan* did not complete until May 1945.

The 'Tribals' saw a considerable amount of action during the Second World War, serving in practically all the theatres of war from the Arctic to the Pacific. Of the 16 original Royal Navy ships 12 were lost in the early years of the war, and no fewer than seven were lost in 1942.

Probably the most famous of the 'Tribals' was the *Cossack* which, on February 17, 1940, violated Norwegian territorial waters by entering Jössing Fiord where the German supply ship *Altmark* lay hidden. This ship carried 299 British seamen from merchantmen captured by the *Graf Spee.* A boarding party from the destroyer seized the *Altmark* and, after a brief fight, released the captives who were transferred to the *Cossack* and brought home. Two months later, on April 13, she took part in the Second Battle of Narvik together with the *Eskimo, Bedouin* and *Punjabi.* During this battle *Cossack* was hit by eight 5-in shells from the damaged destroyer *Diether von Roeder* before she silenced her adversary, which later blew up. Badly damaged, the *Cossack* ran aground where she remained for 13 hours until she freed herself and made her way home. In the meantime the *Eskimo* had sunk the destroyer *Erich Köllner,* with the help of *Bedouin,* and torpedoed the destroyer *Hermann Künne* which had run aground. *Eskimo* was torpedoed by the *Georg Thiele,* her bows from B mounting forward being blown off, but she

HMS *Eskimo* in wartime. The 'Tribals' were designed to lend heavier gun support to existing flotillas and were influenced by similar contemporary German (Type 1934) Japanese (*Asashio* Class) and American (*Gridley* Class) destroyers

IWM

IWM

Left: *Eskimo* in wartime configuration. The superimposed after 4.7-in mounting was replaced by a twin 4-in AA after the Norway campaign, the aft funnel shortened, and a variety of light AA guns added. The later Canadian Tribals were armed with 8 4-in guns. Right: *Tartar* off Iceland

survived and was repaired on the Clyde. Two other 'Tribals' did not survive the Norwegian campaign; on April 9 the *Gurkha* and on May 3 the *Afridi* were sunk by aircraft.

On the night of May 26/27, 1941 the *Cossack* was in action again when, with the *Maori*, *Sikh* and *Zulu*, she carried out several torpedo attacks against the *Bismarck*. Although the destroyers claimed hits on the German battleship they did not in fact score any. On May 28 the *Mashona*, returning home with *Tartar* after being involved in the search for *Bismarck*, was bombed and sunk by German aircraft. The *Cossack* was also lost before the end of the year. On October 23 she was torpedoed by *U 563* west of Gibraltar.

The Mediterranean proved a deadly battleground for destroyers during 1941-43 and among those lost were five of the 'Tribal' Class. The first was the *Mohawk*. On April 16, 1941 in company with the *Nubian*, *Janus* and *Jervis* she attacked an Italian convoy of five merchant ships. All the merchantmen and the three escorting destroyers were sunk by the British force, but the *Mohawk* was hit by two torpedoes from the sinking destroyer *Luca Tariago*. She was quickly abandoned and was later sunk by *Janus*. Better luck accompanied the *Maori* and *Sikh* on the night of December 12/13, 1941. With the destroyers *Legion* and *Isaac Sweers* (Dutch) they intercepted the Italian cruisers *Alberto di Giussano* and *Alberto da Barbiano* off Cape Bon. The *Barbiano* was hit by three torpedoes and sank almost immediately, the *Giussano* was hit by one torpedo amidships, caught fire and sank one hour later. Two months later, on February 12, the *Maori* was bombed and sunk in Grand Harbour, Malta, during an air raid.

On June 15, 1942 the *Bedouin* was damaged in action with Italian naval forces and was later torpedoed and sunk by an aircraft (which she shot down). On September 14, 1942 *Sikh*, *Zulu* and other warships were covering an unsuccessful commando raid on Tobruk. *Sikh* was hit by the shore batteries, disabled and eventually sank; the *Zulu* was bombed and sunk by German aircraft while returning to Alexandria.

Three 'Tribals' were lost in Arctic waters during 1942. The *Matabele* was torpedoed on January 17 by *U 454* in the Barents Sea while escorting PQ8; there were two survivors. On May 1 the *Punjabi* was covering PQ15 in a thick fog when she was accidentally rammed and cut in two by the battleship *King George V*. On September 20, while escorting QP14, the *Somali* was torpedoed by *U 743*. She was taken in tow by *Ashanti* but broke in two and sank four days later.

During 1942-43 the remaining Royal Navy 'Tribals' served in the Mediterranean, covering convoys and the seaborne invasions of North Africa, Sicily and Italy. During the same period the four newly completed Royal Canadian Navy ships served in the Arctic, mainly covering Russian convoys. In 1944 all except *Nubian* joined the 10th Destroyer Flotilla based at Plymouth for operations in the English Channel and Bay of Biscay. On April 27-28 the *Athabaskan (i)* was torpedoed and sunk by the German destroyer *T 24* off St Brieux. *Tartar*, *Eskimo* and *Nubian* ended the war in the Far East while the remainder of the class stayed in home waters.

The surviving Royal Navy ships were placed in reserve shortly after the war ended and were sold for scrap in 1948-49. The Australian and Canadian ships, being newer, survived much longer, and all except *Micmac* served in the Korean war. During the 1950s they were modified as A/S escorts. The *Bataan* was sold for scrap in 1958, and the *Warramunga* in 1963, while *Arunta* foundered in tow to the breakers in 1969. The Royal Canadian Navy vessels were sold for scrap between 1964 and 1970, except for *Haida*, now preserved as a naval museum.

Displacement: 1960 tons *Length:* 114.91 m (377 ft) oa *Beam:* 11.12 m (36 ft 6 in) *Draught:* 3.96 m (13 ft) *Machinery:* 2-shaft geared turbines, 44 000 shp=36 knots *Armament:* 8 4.7-in (120-mm) (4×2); 4 2-pdr pom-poms (1×4); 8 .5-in (12.7-mm) machine-guns (2×4) *Crew:* 190

AG 11

Russian submarine class. In 1916 the Russian navy bought 11 submarines from the American Electric Boat company. They were very similar to the US Navy's 'H' Class, and were known to the Russians as the 'AG' or *Amerikanski Golland* (American Holland) type. Five were allocated to the Baltic *(AG 11-16)* and six to the Black Sea *(AG 21-26)*. A further six were not delivered because of the Revolution, and were purchased for the US Navy as *H 4-9* in 1918.

The *AG 14* was sunk in July 1917, probably by a mine off Libau. Four, *AG 11-13 (AG 13* was renumbered *AG 16*) and *AG 15*, were scuttled in April 1918 to prevent them from falling into German hands. *AG 22* was one of General Wrangel's fleet which chose exile at Bizerta in 1920 rather than life under the new regime, and she was scrapped in the 1930s. *AG 23-26* became the Soviet *A 1-4*; *AG 21* fell into British hands in 1919, was scuttled by them, raised in 1928 and became the Soviet *A 5*. A number of more 'political' names were also borne at various times:
A 1 ex-*No 12*, ex-*Shakter*, ex-*Trotsky*
A 2 ex-*No 13*, ex-*Kommunist*
A 3 ex-*No 14*, ex-*Marxist*
A 4 ex-*No 15*, ex-*Politrobotnik*
A 5 ex-*No 16*, ex-*Metallist*

All five served in the Second World War. *A 1* was lost in mid-1942 near Sevastopol. *A 2* was apparently non-operational by 1943-44 and was scrapped in 1946. *A 3* was sunk on November 4, 1943 off Tendra Island in the Black Sea by a German escort and *A 4* was lost in 1942 or 1943. *A 5* survived the war.

Displacement: 356/434 tons (surfaced/submerged) *Length:* 46 m (150 ft 3 in) oa *Beam:* 4.7 m (15 ft 9 in) *Draught:* 3.7 m (12 ft 6 in) *Machinery:* (surfaced) 2-shaft diesels, 950 bhp=14 knots; (submerged) 2-shaft electric motors, 600 ehp=10.5 knots (12/8 knots in service) *Armament:* 4 45-cm (17.7-in) torpedo tubes (all forward, eight torpedoes carried); 1 45-mm (1.7-in) gun added by 1941 *Crew:* 25

Agano

Japanese cruiser class. Shortly before the Second World War the Japanese navy drew up designs for light cruisers to replace the old vessels dating from the First World War. After several designs were studied it was decided to build a new type of high-speed scout to work with the Combined Fleet.

The four *Agano* Class were unusual in many ways. They were very lightly armoured to ensure maximum speed and carried quadruple 24-in torpedo tubes on the centreline, like destroyers. In addition they were fitted with Sonar and depth-charges, and in many ways they resembled large destroyer leaders more than fleet cruisers.

The fates of the four ships reflect the final destruction of Japanese naval power. The *Agano* was laid down in June 1940 and was completed in October 1942. She was torpedoed by the US submarine *Skate* north of Truk on February 16, 1944. The *Noshiro* was laid down in September 1941 and completed in June 1943, and was sunk by carrier planes from USS *Hornet* and *Wasp* off Panay on October 26, 1944. The *Yahagi* was begun in November 1941 and completed in December 1943. She escorted the giant battleship *Yamato* on her final sortie to Okinawa, but, like her consort, she was sunk by Task Force 58 aircraft 130 miles west-southwest of Kagoshima on April 7, 1945. The fourth ship, *Sakawa*, was laid down in November 1942 but material shortages and delays held up her completion until November 1944. She was surrendered in August 1945 and was expended as a target in the Bikini nuclear test on July 1, 1946.

The antiaircraft armament of all four ships was augmented during the war, increasing to 46 25-mm in 1943. *Yahagi* and *Noshiro* received this armament on completion, and in 1944 they received a further six single 25-mm (1-in) guns. By the July of that year they had 59 25-mm, in addition to two radar sets.

To assist in their scouting role, the *Agano* Class were fitted with a catapult for launching two floatplanes. No hangar was provided as the space was not available, and the floatplanes were stowed on a platform amidships.

Agano launched Sasebo dockyard October 22, 1941
Noshiro launched Yokosuka dockyard July 19, 1942
Sakawa launched Sasebo dockyard April 9, 1944
Yahagi launched Sasebo dockyard October 25, 1942

Displacement: 6652 tons (normal), 8534 tons (full load) *Length:* 176 m (572 ft 6 in) oa *Beam:* 15 m (49 ft 9 in) *Draught:* 5.5 m (18 ft 6 in) *Machinery:* 4-shaft geared steam turbines, 110 000 shp=35 knots *Protection:* 60 mm (2¼ in) belt; 20 mm (¾ in) deck; 25 mm (1 in) turrets; 50 mm (2 in) magazines *Armament:* 6 150-mm (5.9-in) (3×2); 4 75-mm (3-in) AA (2×2); 32 25-mm (1-in) AA (8×3, 8×1); 8 24-in (61-cm) torpedo tubes (2×4); 2 floatplanes *Crew:* 730

Agile

US air-to-air missile. The importance of achieving a 'dog-fight' missile for the engagement of rapidly-manoeuvring targets at any altitude led to the AIM-95 Agile. The concept dates from the second half of the 1960s when the US Naval Weapons Center began a research and development programme aimed at achieving a compact weapon for close-quarter aerial combat, mainly as armament for the F-14 and F-15 air-superiority fighters.

A main task was to achieve an effective all-aspect infrared guidance seeker, for which Hughes Aircraft was awarded a contract in April 1973.

The missile itself, powered by a Thiokol solid-fuel rocket motor, had a cylindrical body with a short tapered section supporting the homing head. Eight small flip-out tail fins served for stability and steering was achieved by gimballing the rocket nozzle.

Although flight trials from a Phantom of

HMS *Agincourt* was ordered by Brazil from Armstrongs, sold to Turkey before she was completed, and seized by Britain on the outbreak of war. She was longer and had more heavy guns (14 12-in), than any other British battleship of the period the US Navy were made in the summer of 1973, the missile was not put into production. Instead, effort was concentrated on achieving a cheaper, lighter weapon and further development of the IR seeker.

A joint services programme sought to combine the results of the USAF's defunct Claw project with the Agile programme. As part of this effort, a US Navy homing head was scheduled for captive tests in early 1977 in competition with an Aeronutronic Ford seeker being developed for the USAF's Armament Development and Test Center.

Length: 2.44 m (8 ft) *Diameter:* 20 cm (8 in) *Launch weight:* Probably less than 136 kg (300 lb)

Agincourt

British battleship. This unusual ship was the result of a prolonged period of rivalry between the chief naval powers of South America, Argentina, Brazil and Chile. When Argentina ordered two ships with 12 12-in (305-mm) guns, the *Moreno* and *Rivadavia*, in 1910, Brazil decided to have the maximum number of guns possible to outclass them. She ordered the *Rio de Janeiro* from Armstrongs on the Tyne, armed with no fewer than 14 12-in guns, the greatest number ever mounted in a battleship. The ship was laid down in September 1911 and launched on January 22, 1913.

To accommodate the seven turrets required a very long hull. The disposition was two twin turrets forward, two amidships between the funnels at forecastle deck level, two at upper-deck level aft and one superimposed. When the ship was in British service the turrets were named after the days of the week instead of the usual 'A', 'B', 'P', 'Q' etc. Another unusual feature was the secondary armament of 20 6-in (152-mm) guns, which was much heavier than any contemporary ship.

Early in 1914 Brazil changed her mind about wanting the *Rio de Janeiro* and put the ship on the market. The rubber boom had collapsed, and in any case the Chileans had ordered ships with 14-in (360-mm) guns, which made the Brazilian giant look less formidable. Turkey bought the ship for £2725000 and renamed her *Sultan Osman I*, but in July 1914 the British government used various pretexts to delay her until it was clear that Britain would be at war. As soon as war broke out the ship was seized and taken into the Royal Navy as HMS *Agincourt*, and although Turkey had already decided to side with the Central Powers, the incident caused great indignation.

The *Agincourt* was altered several times. As soon as she joined the Fleet she lost her hurricane deck between the funnels; in 1916 her tripod mainmast was replaced by a pole mast for a short while, and later her derrick was fitted with a topmast. The ship fought in the 6th Division of the 1st Battle Squadron at the Battle of Jutland. She insisted on firing full 14-gun broadsides to confound the critics who said that her hull could not stand the strain, and the sheet of flame looked to other ships like a battlecruiser blowing up!

The ship was put on the disposal list in 1919 but later began a conversion to a fleet repair ship, which was stopped. She was sold for scrapping in 1922.

See also battleship *Erin*.

Displacement: 27500 tons (normal) 30250 tons (full load) *Length:* 204.5 m (671 ft 6 in) oa *Beam:* 26.8 m (89 ft 0 in) *Draught:* 8.2 m (27 ft) mean *Machinery:* 4-shaft Parsons steam turbines, 34000 shp=22 knots *Protection:* 100-225 mm (4-9 in) belt; 200 mm-300 mm (8 in-12 in) turrets; 25 mm-60 mm (1 in-2½ in) decks; 100 mm-300 mm (4 in-12 in) conning tower. *Armament:* 14 12-in (305-mm) (7×2); 20 6-in (152-mm) (20×1); 10 3-in (76-mm) (10×1); 2 3-in (76-mm) AA (2×1); 3 21-in (53-cm) torpedo tubes (submerged, bow and broadside) *Crew:* 1115 (1267 by 1918)

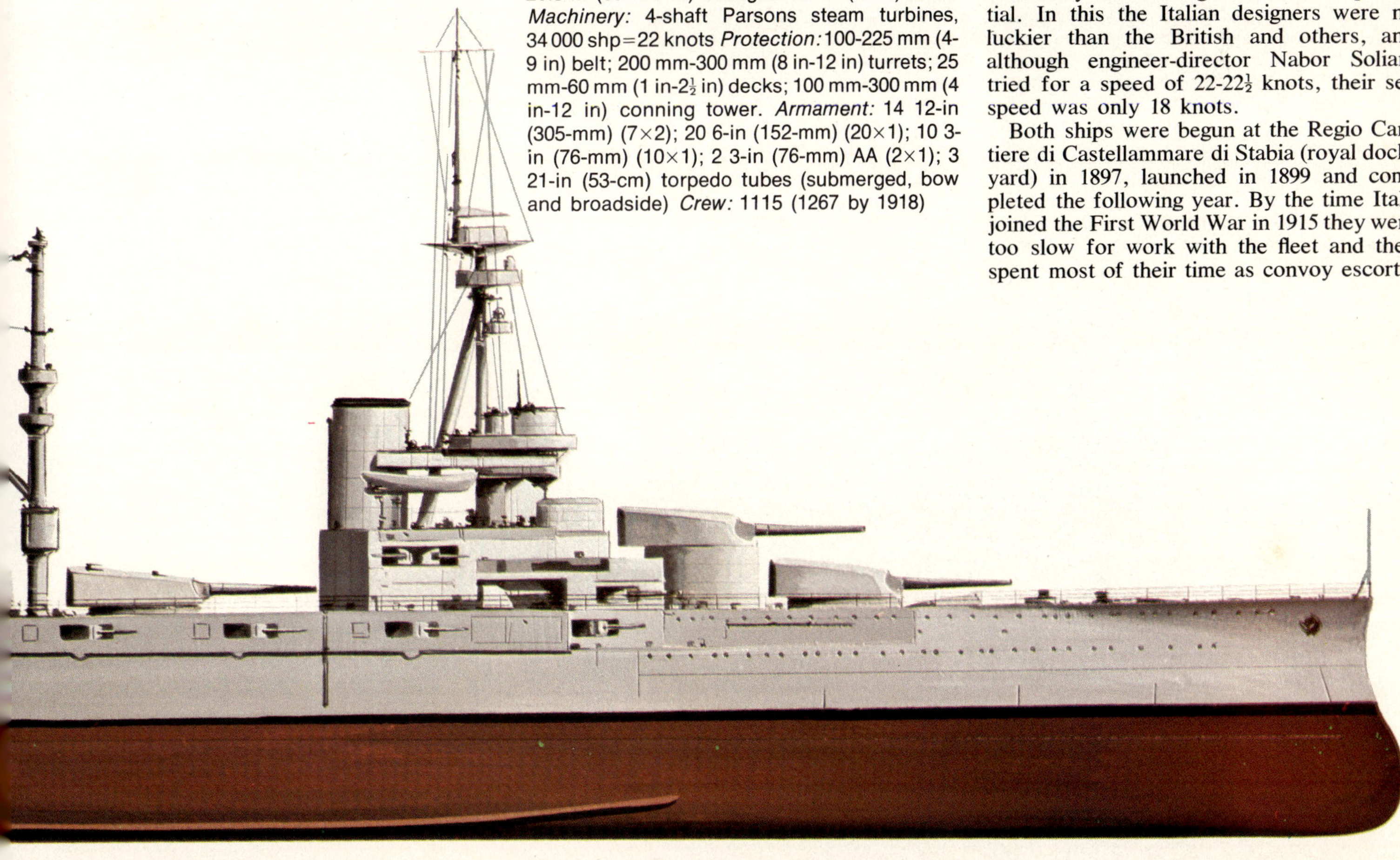

AGM-12 US air-to-surface missile See **Bullpup**

AGM-28 US air-to-surface strategic missile See **Hound Dog**

AGM-45A US air-to-surface antiradar missile See **Strike**

AGM-53 US air-to-surface missile See **Condor**

AGM-62 US air-to-surface guided bomb See **Walleye**

AGM-65 US air-to-surface missile See **Maverick**

AGM-69 US air-to-surface missile See **SRAM**

AGM-78 US air/surface-to-surface anti-radiation missile See **Standard ARM**

AGM-83 US air-to-surface missile See **Bulldog**

AGM-86 US air-launched cruise missile See **ALCM**

AGM-88 US air-to-surface antiradiation missile See **HARM**

Agordat

Italian scout cruiser class. Although rated as scout cruisers, these two small warships corresponded closely to the British category of torpedo gunboats, ie ships intended to run down the early torpedo boats.

The problem for ships of this type was that they were developed at a time when high-speed triple-expansion and compound machinery was nearing the limit of its potential. In this the Italian designers were no luckier than the British and others, and although engineer-director Nabor Soliani tried for a speed of 22-22½ knots, their sea speed was only 18 knots.

Both ships were begun at the Regio Cantiere di Castellammare di Stabia (royal dockyard) in 1897, launched in 1899 and completed the following year. By the time Italy joined the First World War in 1915 they were too slow for work with the fleet and they spent most of their time as convoy escorts.

Although rearmed in 1919-21 with 4.7-in (120-mm) guns fore and aft, they were scrapped shortly afterwards.

Displacement: 1313 tons (normal) *Length:* 91.6 m (300 ft 6 in) oa; 87.6 m (287 ft 5 in) pp *Beam:* 9.32 m (31 ft 2 in) *Draught:* 3.54-3.64 m (11 ft 7 in-11 ft 11 in) *Machinery:* 2-shaft 4-cylinder compound reciprocating, 7500-8000 ihp=22-$22\frac{1}{2}$ knots (designed) *Protection:* 20 mm (0.8 in) deck *Armament:* 12 3-in/40 cal guns; 2 17.7-in (45-cm) torpedo tubes *Crew:* 174-184

Agosta

French submarine class. In 1970 the Marine Nationale (French navy) announced the building of a new class of four high-speed conventionally powered submarines. Although claimed to be of the most advanced design, the details published by mid-1976 had failed to indicate any unusual features. They are presumably intended to maintain continuity of design, and are a logical improvement over the *Daphné* Class of 1958-70.

Published information indicates that there are only four forward torpedo tubes, and no stern tubes, but 20 torpedoes are carried. These are presumably homing torpedoes of the L5 type, which is a 55-cm (21.7-in) weapon with an active/passive head. Two sonar sets are carried, an active DUUA 1 set with transducers forward and aft, and a passive DSUV set with 36 hydrophones.

The class includes four units: the prototype *Agosta*, begun in February 1972 and completed in the spring of 1976; the *Bevéziers*, due for completion about three months later; and the *Ouessant* and *La Praya*. The names commemorate French submarines of the pre-Second World War period, and all four are being built by DCAN Cherbourg (a naval dockyard). Two are reported to be building in Spain at Cartagena, and more are on order for South Africa.

Displacement: 1200 tons (standard); 1450/1725 tons (surfaced/submerged) *Length:* 67.6 m (222 ft) *Beam:* 6.8 m (22 ft) *Draught:* 5.4 m (18 ft) *Machinery:* Single-shaft diesel-electric, 3600 bhp=12 knots (surfaced); single-shaft electric motor, 4600 hp=20 knots (submerged) *Torpedo tubes:* 4 55-cm (21.7-in), 20 torpedoes *Endurance:* 560 (350 miles) at $3\frac{1}{2}$ knots (dived); 14 500 km (9000 miles) at 9 knots (snorting) *Crew:* 50

Agusta, Agusta-Bell Italian helicopters See **Hirundo, Iroquois, Kiowa**

AH-1 US Army attack helicopter See **Hueycobra**

Aichi Japanese aircraft See **D3A**

Aigli

Greek torpedo boat class. These six boats, *Aigli, Alkyone, Arethousa, Dafni, Doris* and *Thetis*, were built in 1912-13 by the German Vulcan firm at Stettin. They resembled early German torpedo boats but were armed with US-pattern guns from Bethlehem Steel, the principal supplier of armaments to the Greeks at this time.

The *Aigli* reached 26.2 knots on trials but the *Doris* only made 25.7 knots. As far as is known, the class did nothing during the First World War. *Dafni* and *Thetis* were sold for scrapping in 1931 but the remaining four were overhauled in 1926-30, losing one torpedo tube. All were lost during the German invasion in April 1941.

Displacement: 120 tons (normal) 145 tons (full load) *Length:* 45 m (147 ft 8 in) *Beam:* 2.84 m (9 ft 4 in) *Draught:* 1.22 m (4 ft) *Machinery:* 2-shaft triple-expansion, 2600 ihp=25 knots *Armament:* 2 6-pdr (57-mm) guns (2×1); 3 (later two) 18-in (46-cm) torpedo tubes (The tubes were originally disposed side by side in the forward well-deck and a single tube on the centreline aft, but during the 1926-30 refit the forward tubes were replaced by one on the centreline. In 1931 *Aigli* had no torpedo tubes at all.) *Crew:* 25 approx

Aigrette

French submarine class. These two submarines were the first in the world to be driven by diesel motors. At the time the German press adopted a most superior tone, hinting that the French were forced to turn to Germany for lack of technical knowhow, but the German navy waited eight years before ordering its first diesel-engined U-Boat. The diesels were supplied by MAN of Augsburg.

A further 11 boats were ordered at the same time, May 13, 1902, but the order was cancelled in September 1902. Their names would have been *Eider, Macreuse, Grèbe, Cygne, Marabout, Héron* (Toulon dockyard), *Pluvier, Pingouin, Pélican, Plongeon* and *Vanneau* (Cherbourg dockyard). The designer was Maxime Laubeuf, and the design was a development of his steam-driven *Narval* and *Sirène* types.

Marius Bar

The *Aigrette*—first diesel-driven submarine

The *Aigrette* (*Q 38*) was launched on January 23, 1904 and completed in 1905, while the *Cicogne* (*Q 39*) was launched on November 11, 1904 and completed in 1908. The *Aigrette* was based at Cherbourg from 1914 to 1918 for local defence, and was experimentally fitted with net-cutters. *Cicogne* was sent to Brindisi for local defence in 1916 and remained there until the Armistice. Both boats were removed from the effective list in November 1919 and scrapped.

Displacement: 178/253 tons (surfaced/submerged) *Length:* 35.85 m (117 ft 7 in) oa *Beam:* 4.05 m (13 ft 3 in) *Draught:* 2.63 m (8 ft 8 in) *Machinery:* Single-shaft diesel, 150 bhp=9.25 knots (surfaced); single-shaft electric motor, 130 hp=6.2 knots (submerged) *Armament:* 2 Drzewiecki drop-collars for 45-cm (17.7-in) torpedoes and two cradles (external) *Endurance:* 2400 km (1300 miles) at 8 knots (surfaced); 120 km (65 miles) at 3.8 knots (submerged) *Crew:* 14

The Bell Airacobra was unique among US fighters of the Second World War period in having its engine mounted in the centre fuselage, driving the propeller via a 10-ft shaft under the cockpit, to enable a 37-mm cannon to fire through the propeller hub. Also standard were 2 .5-in machine-guns mounted in the nose fairing, while the P-39D (shown) had 4 .30-in Brownings in the wings

AIM-4 US air-to-air missile See **Falcon**

AIM-7 US air-to-air missile See **Sparrow**

AIM-9 US air-to-air missile See **Sidewinder**

AIM-26 US air-to-air missile See **Falcon**

AIM-47 US air-to-air missile See **Falcon**

AIM-54 US air-to-air missile See **Phoenix**

AIM-95 US air-to-air dogfight missile See **Genie**

Airacobra P-39 Bell

USAAF fighter. When the Consolidated Company moved from Buffalo to California in 1935, three of its top men, led by Lawrence D Bell, stayed behind to form a new company. From the start it concentrated on fast combat aircraft, building radically unconventional fighters such as the Airacuda, which had two pusher Allison liquid-cooled V-12 engines, with a cannon gunner in the front of each nacelle. The Airacobra had a single Allison—then a completely new engine in a land of air-cooled radials—mounted behind the pilot and driving the propeller through shafting to allow a 37-mm cannon to fire through the hub. The nose also housed two 0.5-in (13-mm) Brownings and the retracted nosewheel, another innovation in October 1937 when the US Army Air Corps ordered one XP-39. This prototype, which flew in

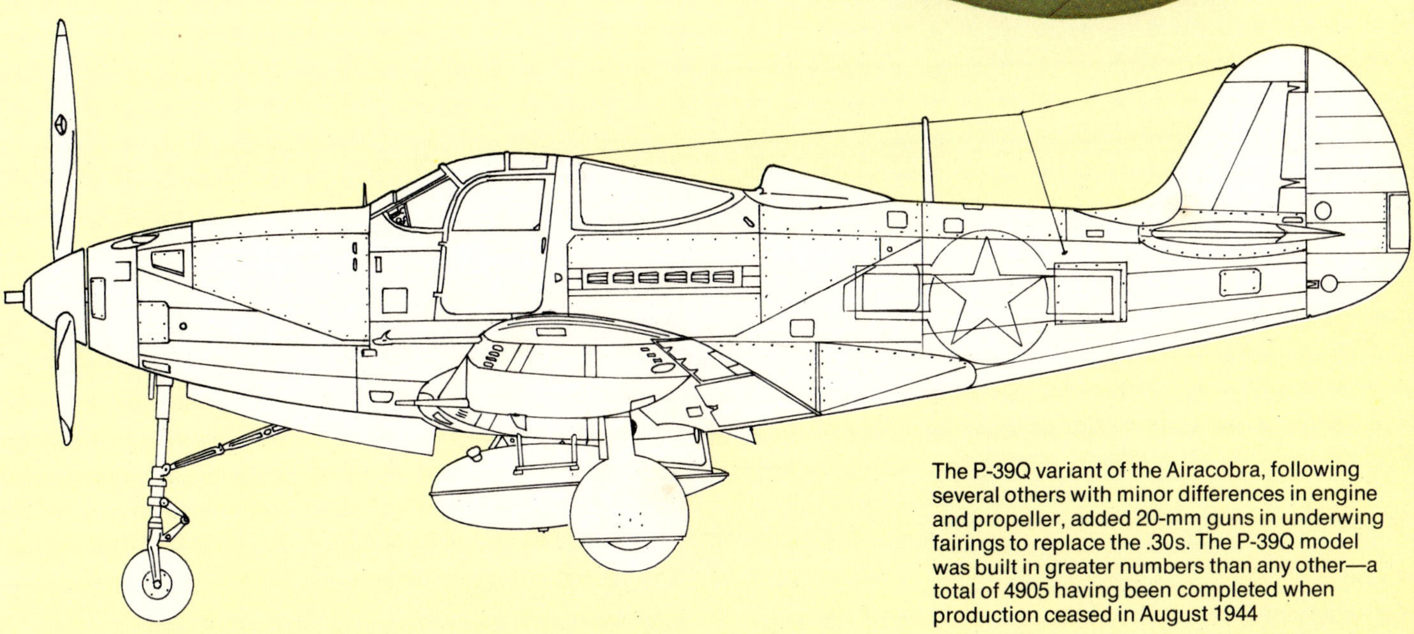

The P-39Q variant of the Airacobra, following several others with minor differences in engine and propeller, added 20-mm guns in underwing fairings to replace the .30s. The P-39Q model was built in greater numbers than any other—a total of 4905 having been completed when production ceased in August 1944

April 1939, had a turbocharged engine and external radiator, but subsequent YP-39s, delivered from September 1940, had the radiator inside the centre section under the engine, which had no turbocharger and made the aircraft poor at altitude. Britain used the first production models to equip 601 Squadron with a 20-mm Hispano, two 0.5-in above the nose and four 0.303-in in the wings. The serviceability was poor, pilots did not like the car-type doors and nosewheel, the RAF disliked the general complexity and strangeness, and the main British batch were given to the Soviet Union or re-equipped for the Army Air Corps with the designation P-400.

After building 336 of these early versions, production was concentrated on the P-39D with 1190-hp Allison V-1710-35 engine, 37-mm gun instead of 20-mm and belly rack for a tank or bomb. Then came the P-39F, K, L and M, with different engines and propellers, more 0.5-in ammunition and other variations; the P-39N, with rear armour instead of bullet-proof rear window and smaller fuel capacity to reduce weight; and the last and most numerous model, the P-39Q, with the four rifle-calibre wing guns replaced by two underslung 0.5-in pods, four-blade propeller and restored fuel capacity. Total production was 9558, of which 4773 went to the Soviet Union under lend-lease. Several hundred were used by the US Army Air Forces in the European theatre, but most US-operated P-39s served in the Pacific and Southeast Asia, the Aleutians, Alaska and the Panama Canal Zone. Although not equal to the P-47 or P-51, the P-39 was a good performer which seldom found the limelight. Once users had accepted the tricycle gear and (for the time) high landing speed, the P-39 was recognized as a tractable and compact fighter/bomber with speed of up to 644 km/h (400 mph) and useful range which at 257 km/h (160 mph) with a drop tank was often 2374 km (1475 miles). Production was terminated in July 1944, when Bell turned over to the P-63 Kingcobra derived from the Airacobra.

Span: 10.4 m (34 ft 0 in) *Length:* 9.2 m (30 ft 2 in) *Gross weight:* (Airacobra I, P-400) 3209 kg (7075 lb); (P-39K) 3810 kg (8400 lb); (P-39Q) 3652 kg (8052 lb) *Max speed:* (up to P-39K) 592 km/h (368 mph); (P-39N, Q) 642 km/h (399 mph)

One of only three XP-59A Airacomets, the first turbojet-powered aircraft built in the USA

Airacomet P-59 Bell

USAF fighter. Though several companies in the United States were working on gas turbines for aircraft propulsion in early 1941, transfer of the British Whittle technology in the middle of that year, before Pearl Harbor, gave them greater impetus, and triggered new engine and airframe development.

The first project was for General Electric at Schenectady to build an Americanized Whittle turbojet (which materialized as the I-A of 4989 kg thrust), while Bell at nearby Buffalo built a fighter to use the revolutionary

Production P-59s adopted a nose armament of one 37-mm cannon and three 0.5-in machine-guns mounted in the nose, in place of the original two 37-mms. Only 50 production machines followed the three prototypes and 13 YP-39 development aircraft

engine. The resulting aircraft was very conservative but troublefree, and though design was not begun until issue of the Army Air Corps contract on September 5, 1941, the first XP-59A Airacomet flew on October 1, 1942, long before any British jet fighter.

Secret work out at the new desert base at Lake Muroc (now Edwards AFB) swiftly perfected the XP-59A as a sweet-flying machine, with laminar-flow wings, two 7257-kg (1600-lb) I-16 engines in odd nacelles under the wing roots, two 37-mm cannon in the nose and speed of 658 km/h (409 mph). This was hardly faster than propeller-driven aircraft but batches were ordered of 13 service-test YP-59A Airacomets and 20 production P-59As powered by the 907.2-kg J31-GE-3. The run was completed in 1943 by 30 P-59Bs with J31-5 engines, increased fuel capacity and other changes, all the A and B models having one 37-mm and three 0.5-in nose guns and wing racks for tanks or bombs. One Airacomet was swapped for an early Meteor and three went to the Navy as XF2L-1s. The type never became operational but complete operational evaluation was assigned to the 412th Fighter Group in 1944.

(P-59A) *Span:* 13.9 m (45 ft 6 in) *Length:* 11.8 m (38 ft 10 in) *Gross weight:* 6214 kg (13 700 lb) *Max speed:* 665 km/h (413 mph)

Airco British aircraft See **Amiens, D.H.1-9**

Airone CRDA Cant Z.506B

Italian general-purpose seaplane. Best-known of the products of the CRDA (Cantieri Riuniti dell' Adriatico) Cant, the Z.506B Airone (Heron) was an outstandingly good three-engined twin-float seaplane which served in many roles throughout the Second World War. Derived from the civil Z.506A in 1936, it was used at first as a reconnaissance machine. The designation Z.506 used the initial of the chief designer, Ing Filippo Zappata, whose predilection for all-wooden airframes was strongly in evidence in the Airone. Despite this it was robust and stood up well to harsh conditions.

Powered by three 750-hp Alfa-Romeo 126

The military Cant Z.506B Airone was a development of the Z.506 commercial transport. Nearly 100 Airones were in service when Italy entered the war, but after carrying out bombing missions on targets in France and the Mediterranean, the type was increasingly relegated to patrol and anti-submarine duties. Unarmed Z.506s served in the postwar Aviazione Militare as rescue and patrol aircraft until 1959

RC34 engines, the Z.506B had a raised cockpit for two pilots in tandem and a long ventral gondola with bomb-aimer, weapon bay and rear-gun position. Usually there was a dorsal turret mounting two 12.7-mm Breda-SAFAT heavy machine-guns. Typical bomb load comprised two 500-kg (1100-lb) bombs, various smaller bombs or a single 53-cm (21-in) torpedo. When Italy entered the war in mid-1940 some 96 were in service as bombers and torpedo attack aircraft but after the Greek islands campaign of 1941 most were used for reconnaissance, convoy escort and air/sea rescue. The Serie XII had new 12.7-mm Scotti guns and a 1200-kg (2645-lb) bomb load and many served the Regia Aeronautica, the Luftwaffe and, after the Italian capitulation, the Co-Belligerent AF—the faction of the Italian air force which served with the Allies after October 1943. Those with the Allies survived in postwar service until 1959.

Span: 26.5 m (86 ft 11 in) *Length:* 19.25 m (63 ft 1 in) *Gross weight:* 12 704 kg (28 008 lb) *Max speed:* 349 km/h (217 mph)

Airspeed British aircraft See **Horsa**

Airtos

Italian air-launched antiship missile. Development of this missile began in 1969 partly to provide a means of countering the growing threat of missile-armed fast patrol boats, hydrofoils and other surface craft having speeds up to about 90 knots.

The weapon has a slim, cylindrical body with a streamlined radome containing an active radar seeker, aft of which is a high-explosive warhead and a solid-fuel rocket motor of 2000 kg (4400 lb) thrust. Control is applied via cruciform wings in conjunction with cruciform tail fins. The prime contractor is Sistel S.p.A.

Airtos has an effective range of 3-11 km (1.8-7 miles) under all-weather conditions of operation, its target being first acquired by the ASV radar of the launch aircraft.

After release it responds automatically, following a pre-programmed vertical flight profile under the control of a radio-altimeter and homing into its target. The 35-kg (77-lb) warhead has an impact or proximity fuze.

Length: 3.9 m (12.8 ft) *Diameter:* 21 cm (8.3 in) *Span:* 86 cm (33.8 in) *Weight:* 191 kg (421 lb)

AJ37 Swedish attack aircraft See **Viggen**

AJ168 Franco-British asm See **Martel**

Ajit HAL-built HS Gnat See **Gnat**

Akagi

Japanese aircraft carrier. After the Washington Naval Disarmament Treaty had been signed in 1922 the Japanese navy was left with six incomplete capital ship hulls which had to be scrapped. The success of the small aircraft carrier *Hosho* and the news that the US Navy intended to convert the battlecruisers *Lexington* and *Saratoga* to aircraft carriers convinced the navy department to press ahead with two conversions of its own.

The *Akagi, Amagi, Atago* and *Takao* had been laid down in 1920-21 as 30-knot battlecruisers of 40 000 tons, armed with ten 16-in guns. All work had stopped in November 1922, and the hulls of *Atago* and *Takao* were broken up on the slip. Work began on converting the *Amagi* in the summer of 1923, but on September 1, 1923 the great Tokyo earthquake shook the dockyard at Yokosuka and inflicted severe damage on her hull. Although work started on her sister *Akagi* at Kure in November 1923, the *Amagi* was too badly strained to be worth repairing, and so she was scrapped and replaced by the hull of the cancelled battleship *Kaga.*

The redesigned ship was launched on April 22, 1925 and completed on March 25, 1927. She had no 'island' superstructure, and two funnels at deck-edge level kept smoke away from the deck. The foremost funnel pointed downwards and outwards, while the second one projected straight up just behind the first. Another bizarre feature was a triple flight deck forward. The lack of superstructure soon proved impractical and a small bridge was added on the starboard side of the flight deck. Full advantage was taken of the Washington Treaty's provisions, and ten 8-in guns were provided, four in twin turrets on the lower flight deck forward and six in broadside casemates right aft, where they were practically useless.

Several changes were made to the *Akagi,* and in May 1937 she was taken in hand for modernization. The triple flight deck was removed to allow the main flight deck to be extended to the bows. The 8-in turrets were removed, although surprisingly the casemate positions were left in place. The simplified hangar arrangements allowed the aircraft complement to be raised from 61 to 91. The twin funnels were replaced by a single trunked funnel which curved downwards and slightly aft.

A proper island was provided, but sited on the *port* side. The idea was partly to improve pilots' view while landing, and partly to provide a starboard landing circuit to avoid congestion when operating with her near-sister *Kaga.* Unfortunately these ideas proved misguided, and caused far more deck-landing accidents than before.

At the outbreak of war the *Akagi* was the flagship of Vice-Admiral Nagumo, commanding Carrier Division I, and she led the attack on Pearl Harbor. During January 20-22, 1942 her aircraft attacked Rabaul, Kavieng, Lae and Salamua, and on February 19 they made a devastating raid on Darwin in northern Australia. She covered the invasion of Java in February and March 1942 and in April 1942 attacked Colombo. During the raids in the Indian Ocean *Akagi*'s air group inflicted great damage on the British, including the sinking of the carrier *Hermes.*

At the Battle of Midway on June 3-6, 1942 Carrier Division I was part of the First Mobile Force. *Akagi* carried 21 Mitsubishi A6M2 Zero fighters, 21 Aichi D3A1 'Val' dive-bombers and 21 Nakajima B5N2 'Kate' torpedo-bombers.

On June 4 *Akagi* attacked Midway Island's defences, and was slightly damaged by a shore-based TBF Avenger bomber which bounced off her flight deck at about 0707. But the next American attack, by VS 6 from USS

US Navy

***Akagi* (below) after being refitted with a single flight deck in 1936-38, and (above) in the summer of 1941, with three Zeros forward and her recognition signal aft on the flight deck**

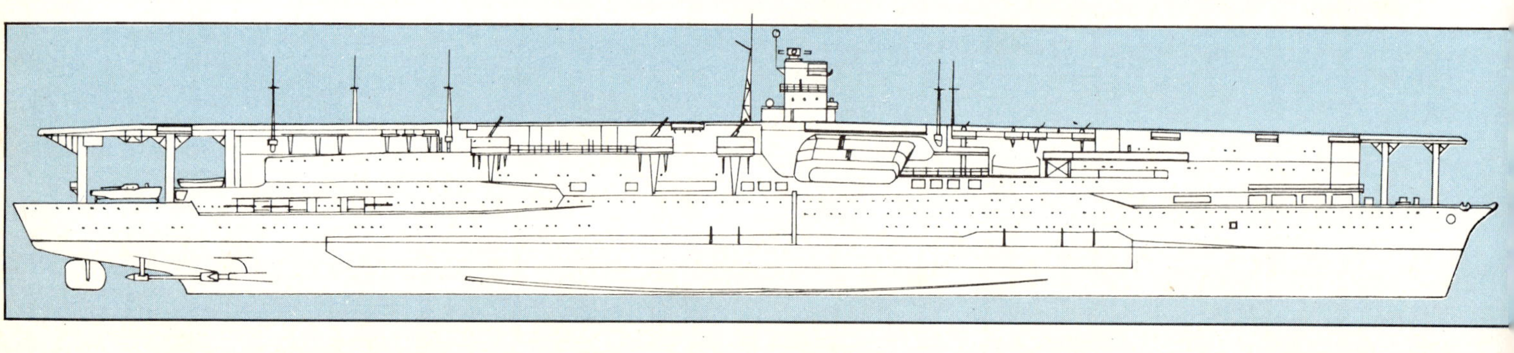

Enterprise, at 1022, caused great damage. She was near-missed by ten yards and then hit twice. A 1000-lb bomb detonated in the hangar and set off torpedo warheads and a gasoline fire, while a 500-lb bomb set fire to aircraft on the flight deck.

By 1047 Admiral Nagumo had to shift his flag to the light cruiser *Nagara,* leaving *Akagi* to burn for another nine hours or more. Admiral Yamamoto hoped to save her, but when he decided to withdraw at dawn on June 5 she had to be abandoned. Four destroyers torpedoed her with 'Long Lance' torpedoes. Surprisingly, her casualties were light, only 263 dead and missing, the lowest losses of any of the carriers at Midway.

(As built) *Displacement:* 26 900 tons (standard) *Length:* 260.6 m (855 ft 4 in) oa *Beam:* 31.3 m (95 ft) *Draught:* 8.07 m (26 ft 6 in) mean *Machinery:* 4-shaft turbines, 131 000 shp=31 knots *Protection:* Belt and hangar deck armour, thickness unknown *Guns:* 10 8-in (204-mm) (6×1, 2×2), 12 4.7-in (140-mm) AA (6×2) *Aircraft:* 61 *Crew:* 1200 approx

(As reconstructed) *Displacement:* 36 500 tons (standard) *Length:* As before *Beam:* (102 ft 9 in) *Draught:* 8.6 m (28 ft 6 in) mean *Machinery:* 4-shaft geared turbines, 133 000 shp=31 knots (oil-fired boilers in place of mixed-firing) *Protection:* Similar to original scheme, unknown *Guns:* 6 8-in. (6×1); 12 4.7-in AA (6×2); 28 25-mm AA (14×2) *Aircraft:* 91 *Crew:* 1340

Akashi

Japanese protected cruiser class. The *Akashi* and her sister *Suma* were the first cruisers built to a Japanese design and by Japanese workmen, at Yokosuka dockyard in 1892-99. The design closely resembled British 2nd Class cruisers of the period, with 6-in guns fore and aft and 4.7-in guns in sponsoned positions in the waist. As usual the guns were supplied by Armstrong, Whitworth.

Marius Bar phot. Toulon

***Suma,* sister ship of *Akashi.* They were the first two cruisers built to a Japanese design**

The *Suma* was found to be deficient in stability and so the *Akashi* was altered during construction, with more freeboard amidships and no fighting tops. At a later date both ships had their mainmasts removed and the *Suma* was reboilered with water-tube boilers of Japanese Miyabara design. *Suma* was completed in 1896 and *Akashi* in 1899.

Both ships fought in the Russo-Japanese war. *Akashi* fought at Chemulpo and both ships were in action at the Battle of the Yellow Sea. *Akashi* was damaged by a mine on December 10, 1904 and had to be repaired at Sasebo. At Tsushima the *Akashi* fought in the 4th Cruiser Division while *Suma* was in the 6th Cruiser Division. Both ships were disarmed in 1922 and sold for scrap in 1927.

Displacement: 2657-2756 tons (normal): *Length:* 89.99 m (295 ft 3 in) pp; *Suma* 93.5 m (306 ft 9 in) *Beam:* 12.72 m (41 ft 9 in); *Suma* 12.19 m (40 ft) *Draught:* 4.80 m (15 ft 9 in) mean; *Suma* 4.65 m (15 ft 3 in) *Machinery:* Two-shaft triple-expansion, 7600 ihp=$19\frac{1}{2}$ knots *Protection:* 25-50 mm (1-2 in) deck, 115 mm (4.5 in) gunshields *Armament:* 2 6-in (2×1); 6 4.7-in (6×1); 10 3-pdr (47-mm) (10×1); 4 $2\frac{1}{2}$-pdr (42-mm) (4×1); 4 Maxim machine-guns; 2 14-in (35-cm) torpedo tubes (above water, broadside) *Crew:* 310

Akatsuki

Japanese destroyer class. Two destroyers were ordered under the 1898 Programme from the Clyde firm of Yarrow, based on the same company's *Ikadzuchi* design ordered three years earlier. They resembled the contemporary British '30-knotter' type destroyers in virtually all respects but carried an extra 12-pdr gun, as the Japanese navy always stipulated heavy armament for its torpedo-craft.

The two ships were ordered in November 1900 and launched and completed in 1901-02. At the start of the Russo-Japanese war they took part in the surprise attack on Port Arthur on February 9, 1904, and a torpedo from the *Kasumi* was claimed to have hit the Russian cruiser *Pallada.* The *Akatsuki* was hit twice during the action.

On May 17, 1904 she was mined and sank off Cape Liao Ti Chan. The *Kasumi* was stricken from the effective list in 1912 and served as a target for some years. She was scrapped about 1920.

Displacement: 363 tons (normal) *Length:* 68.43 m (224 ft 6 in) oa *Beam:* 6.25 m (20 ft 6 in) *Draught:* 1.98 m (6 ft 6 in) max *Machinery:* 2-shaft vertical triple-expansion, 6500 ihp=31 knots *Armament:* 2 12-pdr (2×1); 4 6-pdr (4×1); 2 18-in (46-cm) torpedo tubes (2×1) *Crew:* 59

IWM

The first *Akatsuki* (above) was sunk during the Russo-Japanese war. *Ikadzuchi* (below), of the second *Akatsuki* Class, was sunk in April 1944 by an American submarine

PPL

Akatsuki

Japanese destroyer class. Group III of the 'Special Type' destroyers was ordered under the 1927 New Reinforcement Programme. The four ships were also known as the 'Modified *Fubuki*' Class, and they differed from the first two groups of the 'Special Type' in their machinery.

Improvements in boiler design meant that the new class used three boilers instead of four to develop the same horsepower and speed. With only one boiler in the forward boiler-room it was possible to give them a thinner forefunnel, which gave them a distinctive look. The opportunity was also taken to enlarge and improve the bridge, with a new fire-control system. All four were completed with shields to their torpedo tubes and reloads were carried for the two after banks of torpedo tubes. The *Hibiki* was the first all-welded ship in the Imperial Japanese Navy.

In 1936 all four vessels were altered to improve stability, with strengthened hulls and extra ballast. This raised their tonnage and reduced top speed to 34 knots. Between 1941 and 1943 the light armament was increased to 14 25-mm antiaircraft guns, with two triple mountings in place of 'X' 5-in turret. In 1944 the *Hibiki* was given a total of 22 25-mm guns (7×1, 3×2, 3×3) and ten 13-mm machine-guns (4×2, 2×1). Type 13 and Type 22 radars were installed. In all four the minesweeping gear was replaced by four depth-charge throwers and 36 depth charges.

Akatsuki was sunk by gunfire from the US cruiser *Atlanta* off Savo Island on November 13, 1942. The *Ikadzuchi* was torpedoed by the US submarine *Harder* 200 miles south-southeast of Guam on May 14, 1944 and the *Inadzuma* was torpedoed a month later off Tawi Tawi by the USS *Bonefish*. Only the *Hibiki* survived and she was handed over to the Soviet Union as reparations in July 1947. Renamed *Prytkty*, she served for some years in the Far East and was scrapped in the 1950s.

Displacement: 1680 tons (standard) 1980 tons (normal after modification) *Length:* 118.49 m (388 ft 9 in) oa *Beam:* 10.36 m (34 ft. 0 in.) *Draught:* 3.28 m (10 ft 9 in) *Machinery:* 2-shaft geared steam turbines, 50 000 shp=8 knots (34 knots after modification, sea speed about 30 knots) *Armament:* 6 130-mm (5-in) dual-purpose (3×2); 2 13-mm (0.5-in) AA; 9 61-cm (24-in) torpedo tubes (18 torpedoes carried) *Crew:* 200

US Navy

Akitsuki, shown here in her original form, was the first of a class of destroyers completed from 1942 on, and designed to provide antiaircraft screening for carriers. It was then decided to equip them for the fleet escort as well as the AA role, and to their original main armament of twin quick-firing 100-mm high velocity AA guns were added a quadruple torpedo tube and six depth-charge throwers. Subsequently, the light antiaircraft armament was greatly increased. *Akitsuki* was sunk in October 1944 during the Battle of Leyte Gulf

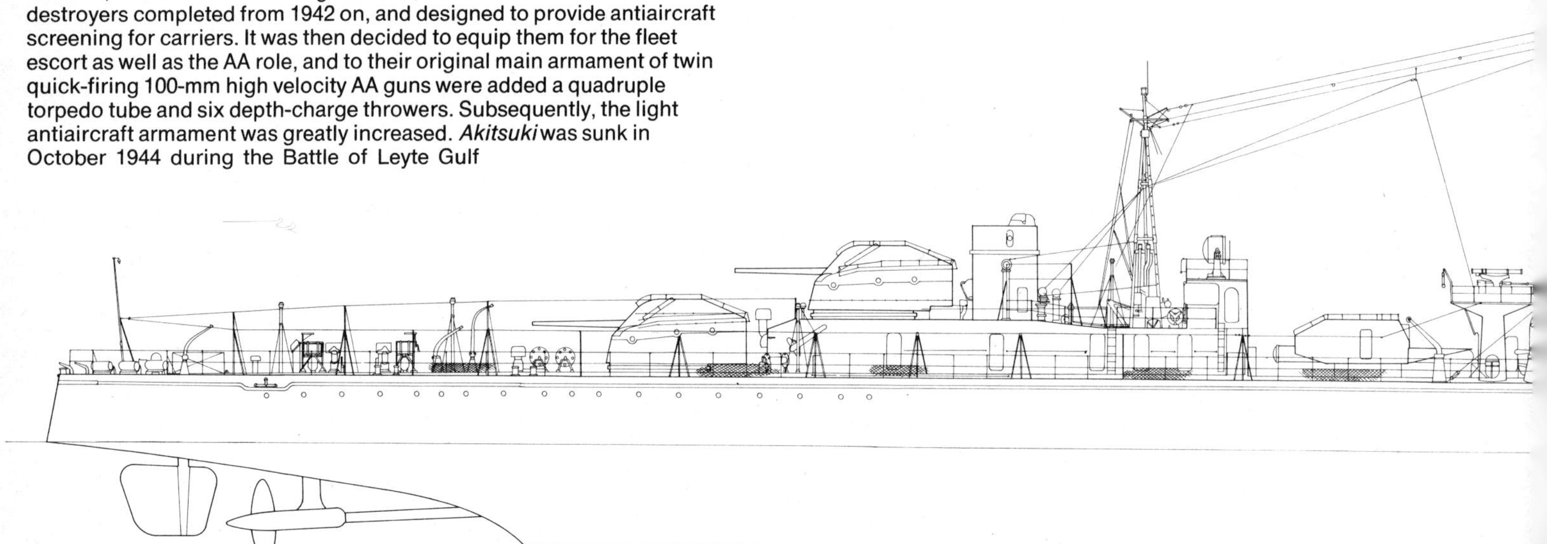

The Japanese battleship *Aki* was the first Japanese turbine-engined battleship. She is shown (left) as completed with torpedo nets fitted, and (below) after being disarmed for use as a target ship in 1923. She was sunk by aircraft bombs in September 1924

Aki

Japanese battleship. This ship was a modified version of the 'semi-dreadnought' *Satsuma* of 1909. During construction the design was considerably modified, with steam turbines in place of reciprocating machinery and a different arrangement of armour.

National Maritime Museum

Japan was only the second nation in the world to build a turbine-engined battleship, and no facilities existed in the country for building such advanced machinery. Brown-Curtis turbines had to be ordered from the US, which combined with financial problems to delay completion for some years.

The *Aki* was laid down at Kure dockyard in March 1906, launched on April 15, 1907 but not completed until March 1911. She had an uneventful career even in the First World War, and was disarmed to comply with the Washington Treaty. She was converted at Kure in 1923 for service as a target, and was used in tests of the Type 93 'Long Lance' torpedo. She was finally sunk by aircraft bombs in 1924.

Displacement: 19 800 tons (normal) *Length:* 152.10 m (499 ft 0 in) oa *Beam:* 25.45 m (83 ft 6 in) *Draught:* 8.76 m (28 ft 9 in) *Machinery:* 2-shaft steam turbines, 24 000 shp=20 knots *Protection:* 230-100 mm (9-4 in) belt; 50 mm (2 in) deck; 200-250 mm (8-9¾ in) barbettes; 250 mm (10 in) turrets *Armament:* 4 300-mm (12-in) (2×2); 12 250-mm (10-in) (6×2); 8 150-mm (6-in) (8×1); 12 75-mm (3-in) (12×1); 4 machine-guns; 5 45-cm (17.7-in) torpedo tubes *Crew:* 931

Akitsuki

Japanese destroyer class. In the late 1930s the Imperial Japanese Navy gave serious thought to the problem of defending the fleet against air attack. The Type B destroyers evolved out of a requirement for a light cruiser, and in size they certainly approached the cruiser category.

The main element of the design was a new twin 100-mm/65-cal high-velocity antiaircraft gun mounting. With its high rate of fire and a range of 18 300 m (20 000 yds), it was more powerful than the contemporary 5-in/38-cal US Navy gun. To increase its effectiveness, the new design was given a second fire-control position aft. To add an offensive capability, the design was subsequently altered to include a bank of quadruple 24-in torpedo tubes amidships.

By re-arranging the boiler- and engine-rooms the uptakes were trunked into one large funnel. This made the ships look remarkably like the light cruiser *Yubari*, and for some months in 1942, when the first vessels appeared, American intelligence reported the *Yubari* appearing in several places at once. Six ships were ordered (Nos 104-109) under the 1939 Programme and a further ten under the 1941 Programme. (Nos 360-369). During the war Nos 366-369 were cancelled for lack of materials and additional units numbered 770-785 and 5061-5083 were cancelled before being laid down. The following ships were built:

Maizuru Dockyard
Akitsuki, Fuyutsuki, Hanatsuki, Hatsutsuki, Hazuki, Kiyotsuki

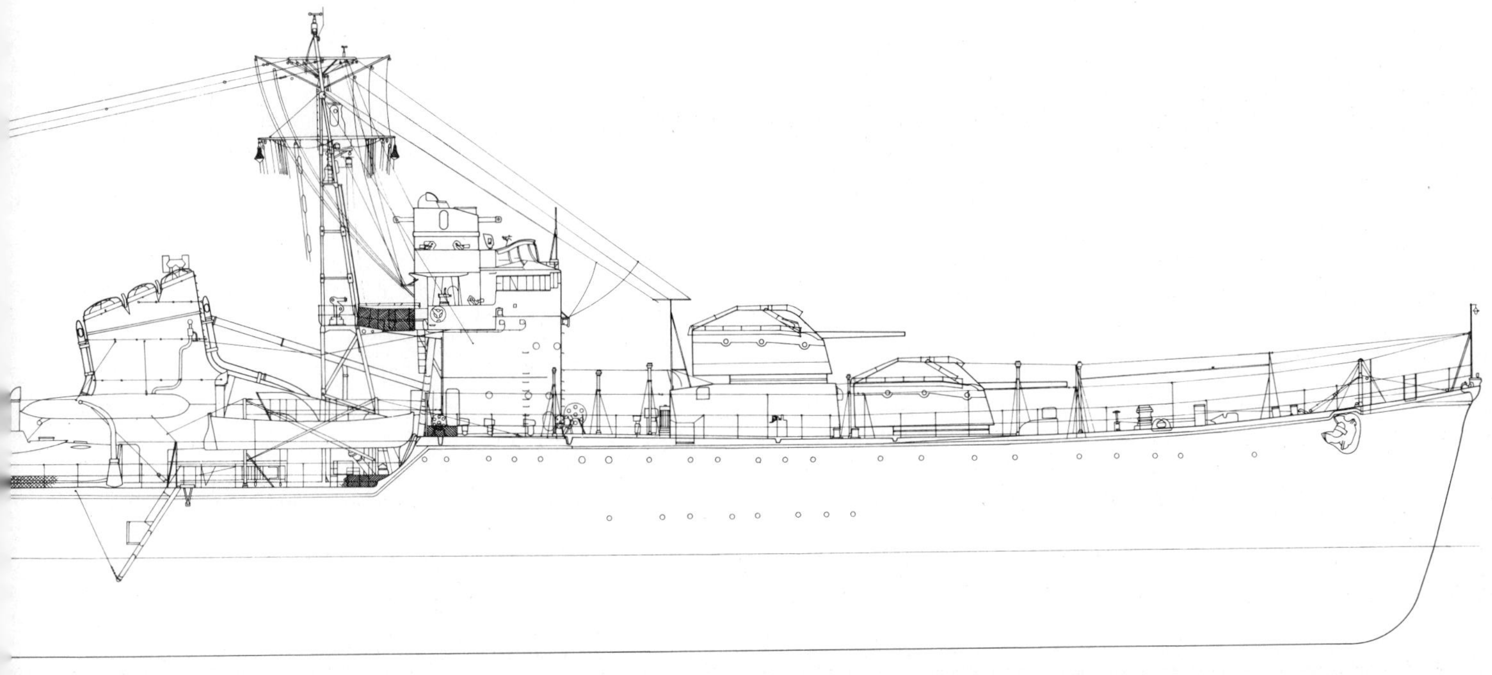

Sasebo Dockyard
Harutsuki, Michitsuki, Natsutsuki

Mitsubishi, Nagasaki
Niitsuki, Ootsuki, Shimotsuki, Suzutsuki, Terutsuki, Wakastsuki

Uraga Dock Co.
Yoitsuki

The lead ship, *Akitsuki*, was started in July 1940 and completed in June 1942, but the last of the class was not completed until April 1945 *(Hatsutsuki)*. The *Hatsuki, Kiyotsuki* and *Ootsuki* were never laid down and the *Michitsuki* was stopped in March 1945 and broken up to make way for suicide craft. Six were sunk in action or torpedoed, and the other six were surrendered, two of them badly damaged. The *Fuyutsuki* was taken over by the US Navy in 1947 and scrapped, the *Harutsuki* became the Soviet *Pospechny*, the *Hatsutsuki* went to Britain and the *Yoitsuki* became the Chinese *Fen Yang*.

Soon after completion the minesweeping gear was replaced by four depth-charge throwers and 54 depth-charges to cope with the growing menace of Allied submarines. Later the outfit was increased to 72 depth-charges. The light armament was soon increased to 15 25-mm AA guns and Type 21 and Type 22 radars were fitted. In June 1944 the typical armament was 29 25-mm guns and 4 13-mm, but the later ships were completed with 40 to 51 25-mm (5/7×3, 25/30×1).

Displacement: 2701 tons (normal) *Length:* 134.21 m (440 ft 4 in) oa *Beam:* 11.58 m (38 ft 0 in) *Draught:* 4.11 m (13 ft 6 in) mean *Machinery:* 2-shaft geared turbines, 52000 shp=33 knots *Armament:* 8 100-mm (4-in) dual-purpose (4×2); 4 25-mm (1-in) AA (2×2); 4 60-cm (24-in) torpedo tubes (8 torpedoes) *Crew:* 290

Akitsushima

Japanese protected cruiser. Built in Japan from imported material, this protected (ie armoured only with an arched 'protective' deck) cruiser was a reduced version of the USS *Baltimore*. Like that ship, the design was drawn up by Sir William White of Armstrong, Whitworth at Elswick-on-Tyne.

The ship was something of a disappointment in service, being rather slow and a bad roller. The arrangement of armament was unusual, with the 6-in guns in sponsons on the beam and 4.7-in guns in shields on the poop and forecastle, and in sponsons between the 6-in.

The *Akitsushima* formed part of the Flying Squadron at the Battle of the Yalu River on September 17, 1894 during the Sino-Japanese war. After the battle she and the cruiser *Naniwa* sank the small Chinese cruiser *Kuang Chi* at Dalny. She also served in the Russo-Japanese war and was part of the scouting force which shadowed the Russian fleet before the Battle of Tsushima. She was removed from the effective list in 1921 and scrapped in 1923.

Displacement: 3100 tons (normal) *Length:* 91.74 m (301 ft 0 in) pp *Beam:* 13.11 m (43 ft 0 in) *Draught:* 5.64 m (18 ft 6 in) max *Machinery:* 2-shaft horizontal triple-expansion, 8400 ihp=19 knots (16 normal max) *Protection:* 75 mm (3 in) deck; 115 mm ($4\frac{1}{2}$ in) gunshields *Armament:* 4 150-mm (6-in) (4×1); 6 120-mm (4.7-in) (6×1); 8 57-mm (3-pdr) (8×1); 4 37-cm (14-in) torpedo tubes (4×1, above water) *Crew:* 330

Akula

Russian submarine. This boat, built in 1906-11, was considered by the Imperial Russian Navy to be their most successful pre-1914 design. She had an unusually heavy armament for her day, two bow torpedo tubes, two stern tubes and four torpedoes carried externally in drop collars.

The *Akula* served in the 1st Division of the Baltic Fleet. In 1915 she was based on Reval and operated with British submarines under Commander Max Horton. Although she was one of the more active Russian submarines she achieved no successes. She left Reval on November 27, 1915 to lay four mines south of Libau; she was reported off Ösel the following evening but that was the last time she was seen. She probably ran into a German minefield northwest of Lyserort.

Displacement: 370/468 tons (surfaced/submerged) *Length:* 38.7 m (187 ft) *Beam:* 3.6 m (12 ft) *Draught:* 3.3 m (11 ft) *Machinery:* 3-shaft diesels, 900 bhp=10.65 knots (surfaced); 3-shaft electric motors, 300 ehp=6.39 knots (submerged) *Armament:* 1 47-mm (1.8-in) gun (added 1914-15); 8 torpedo tubes, probably 45-cm (17.7-in) (4 internal, 4 external) *Crew:* unknown

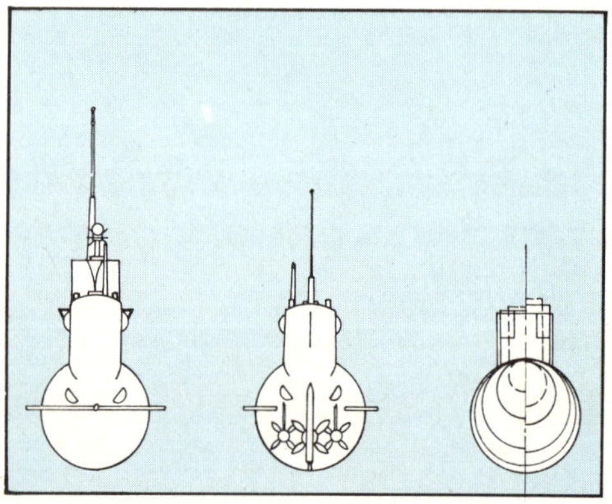

The Russian submarine *Akula* showing, left to right, the view from the bow looking aft, from the stern looking forward, and the body plan

The *Akula* was considered by the Russians to be their most successful design before 1914. Heavily armed with two bow and two stem torpedo tubes and four more torpedoes carried externally in drop collars, and driven by three screws, she had an active career in the Baltic during the first two years of the First World War. She was lost in the Baltic in November 1915 on a minelaying expedition

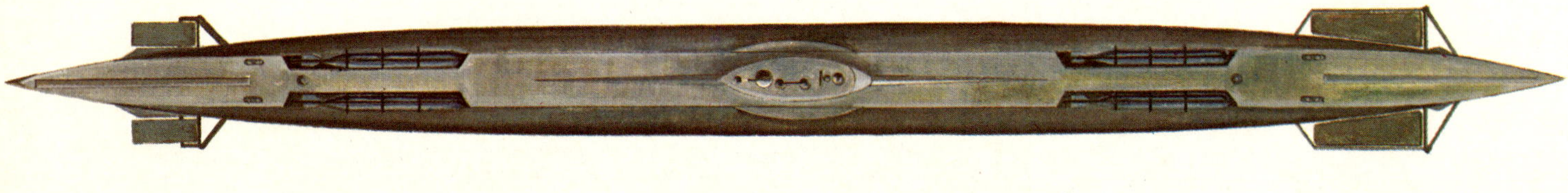

Alabama

US battleship. *Alabama* was the last of the four *South Dakota* Class battleships to be completed. They were 15.25 m (50 ft) shorter and 0.9 m (3 ft) deeper than the preceding *North Carolinas*, which gave them better protection over a greater proportion of their length for virtually the same armament.

Legend credits the *South Dakotas* with 406-mm (16-in) armour, but in fact they were only moderately armoured by contemporary standards. The main belt amidships (which was recessed, giving a smooth-sided hull) was only 310 mm (12.2 in) thick.

Alabama was built very quickly. Laid down at Norfolk naval yard in February 1940, she was launched in February 1942, and commissioned in August of the same year at a cost of over $77 million.

In early 1943 she was attached to the British Home Fleet, escorting North Atlantic and Russian convoys. In August 1943 she was transferred to the Pacific, where she spent the rest of the war with the other fast American battleships acting as escorts to the fast carriers.

Alabama took part in most of the raids against the Japanese-held Pacific islands, and was present at the 'Battle of the Philippine Sea' in June 1944 and the battle for Leyte Gulf in October of that year. However, she took no part in any of the surface actions, her main function being that of heavy antiaircraft escort.

She went into reserve at Bremerton after

USS *Alabama* in December 1942. The *Alabama* served with the British Home Fleet during the first half of 1943 on convoy escort duties in the North Atlantic and Arctic, before transferring to the Pacific to act as heavy antiaircraft escort to the carrier fleets

US Navy

***Alabama* in August 1943 on her way to join the other fast American battleships in the Pacific**

US Navy

Sixteen-inch gun practice on the *Alabama*, one of the smaller and better protected, but equally heavily armed—three triple 16-in turrets were mounted—derivatives of the *North Carolina* Class

decommissioning in January 1947, and was handed over to the State of Alabama in June 1964. Since September 1964 she has been moored as a national memorial near Mobile.

Displacement: 39 000 tons (standard), 44 500 tons (full load) *Length:* 207.1 m (679 ft 5 in) oa *Beam:* 33.3 m (108 ft 1½ in) *Draught:* 10.47 m (36 ft 2 in) *Machinery:* 4-shaft geared steam turbines, 130 000 shp=27.8 knots *Protection:* 310 mm (12.2 in) inclined belt; 146 mm (5.75 in) main deck; 457 mm (18 in) turret faces *Armament:* 9 16-in (406-mm); 20 5-in (130-mm); 48 40-mm; 56 20-mm; *Aircraft:* 2 *Crew:* 2332

Alaska

US battlecruiser class. Although always regarded as battlecruisers, the official designation of these unusual ships was 'CB' or large cruiser. The original class of six ships was conceived as a reply to the German 'pocket battleships' and the imaginary *Chichibu* Class which US Naval Intelligence then believed the Japanese to be building.

The ships that resulted were remarkable in many ways and, although hardly cost-effective, a great credit to the American shipbuilding industry. They were armed with a completely new mark of 12-in (304-mm) gun, and steamed at 31 knots. The armour was only 9 in (127 mm) on the belt, and the scheme of distribution stamped the design as a cruiser rather than a battleship. The machinery of the *Essex* Class fleet carriers was duplicated to save time, but unlike the carriers the *Alaska* Class were not economical steamers. The cost of the *Alaska* was $67 million, of which the hull and machinery cost $45.6 million.

The USS *Alaska* was launched at Camden, New Jersey on August 15, 1943 and commissioned on June 17, 1944. In January 1945 she left for Pearl Harbor, and joined Task Group 58.5 at Ulithi Atoll in the Western Carolines in February. She operated with the fast carriers *Saratoga* and *Enterprise*, screening them on the first night strikes against Tokyo and its airfields. In March she screened the forces attacking Iwo Jima, as part of Task Force 58. She shot down two *kamikaze* bombers on March 18 and the next day had to escort the badly damaged carrier *Franklin* clear of the battle zone. At the close of the

USS *Alabama*. The *South Dakota* Class battleships, of which *Alabama* was the fourth and last, were smaller versions of the preceding *North Carolina* Class. Reduced in length by some 50 ft, they were given much better protection while mounting virtually the same main armament. Large underwater bulges and internal bulkheads gave protection against torpedoes, and the main belt was 12¼ in thick. Note the extensive light AA armament of 48 40-mm and 56 20-mm guns

Okinawa campaign TF 58 attacked shipping in the East China Sea.

The *Alaska* returned home in November 1945 and was decommissioned in February 1947. Her sister *Guam* was launched on November 12, 1943 at Camden, New Jersey by the New York Shipbuilding corporation and commissioned in September 1944. She joined the *Alaska* at Ulithi in March 1945 and screened the fast carriers. She carried out various bombardments between March and June in the Okinawa campaign, and remained at sea for three months. In June 1945 she became the flagship of Cruiser Task Force 95, with the *Alaska*, four light cruisers and nine destroyers. The *Guam* returned to Bayonne, New Jersey in December 1945 and decommissioned there on February 17, 1947.

The third ship, *Hawaii* (CB 3), was not launched by the time the war ended. The launch was postponed several times, but she finally took the water on November 3, 1945 and was laid up in reserve. In 1952 she was reclassified as a large Tactical Command ship (CBC 1) but no work was done; subsequent discussions centred around her conversion to a guided missile cruiser but in September 1954 she reverted to CB 3. She was stricken in 1958 and sold a year later for scrapping.

The *Philippines*, *Puerto Rico* and *Samoa* (CB 4-6) were all authorized in July 1940 and ordered two months later, but the steel shortage of 1943 resulted in the contracts being cancelled in June 1943.

Neither *Alaska* nor *Guam* ever saw service again, as both were stricken from the Navy List in June 1960 and sold for scrapping. They had proved expensive hybrids which did not fit into the US Navy organization. They did very little that smaller cruisers could not have done just as well, and lacked the armour to face bigger ships.

Displacement: 31 700 tons (normal), 34 250 tons (full load) *Length:* 246.4 m (808 ft 6 in) *Beam:* 27.8 m (91 ft 1 in) *Draught:* 9.9 m (32 ft 4 in) max *Machinery:* 4-shaft double-reduction geared turbines, 150 000 shp=33 knots (designed) 31 knots (sea speed) *Protection:* 230 mm (9 in) belt; 80-120 mm ($3\frac{1}{4}$-$4\frac{3}{4}$ in) decks; 320 mm ($12\frac{3}{4}$ in) turret face *Armament:* 9 300-mm (12-in) (3×3); 12 130-mm (5-in) dual-purpose (6×2); 56 40-mm (1.6-in) Bofors AA (14×4); 34 20-mm (0.79-in) Oerlikon AA (34×1) *Crew:* 1773-2251

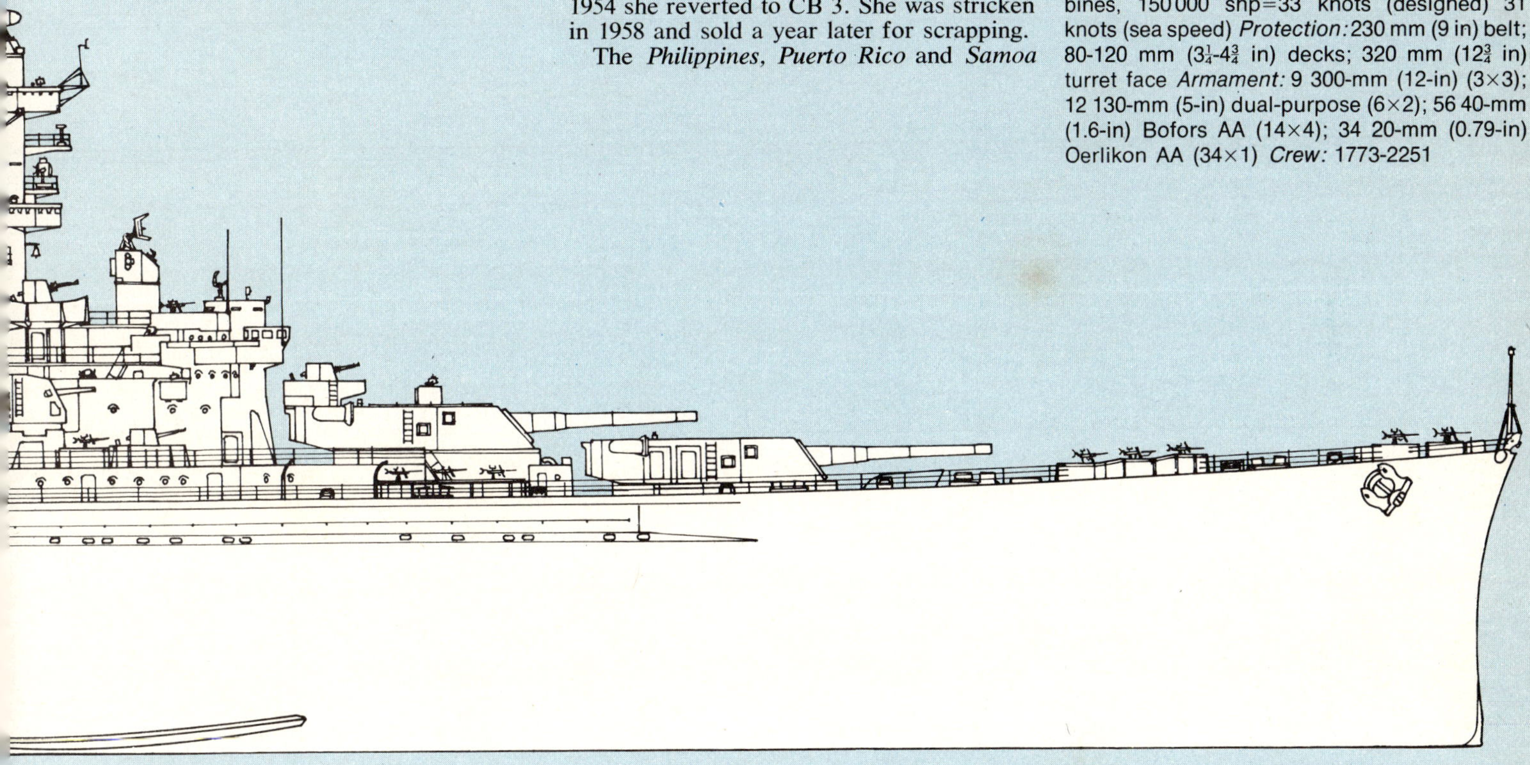

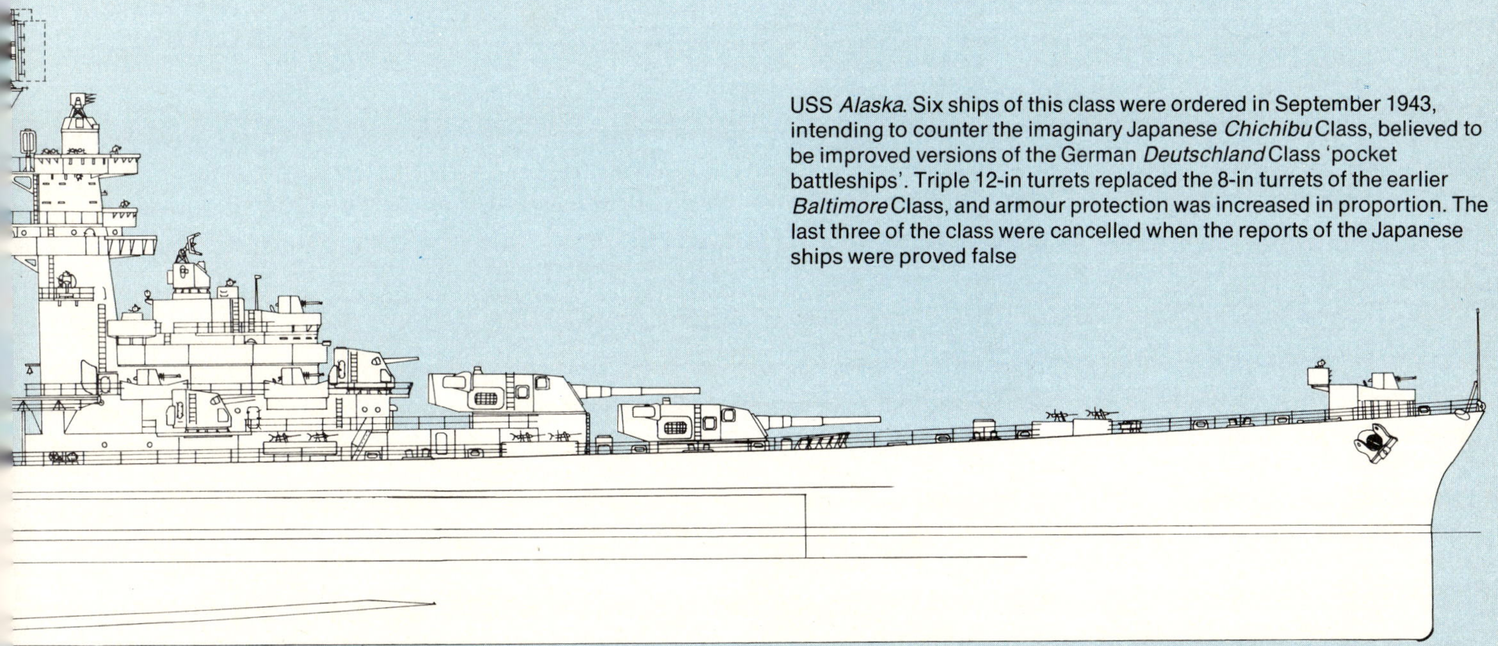

USS *Alaska*. Six ships of this class were ordered in September 1943, intending to counter the imaginary Japanese *Chichibu* Class, believed to be improved versions of the German *Deutschland* Class 'pocket battleships'. Triple 12-in turrets replaced the 8-in turrets of the earlier *Baltimore* Class, and armour protection was increased in proportion. The last three of the class were cancelled when the reports of the Japanese ships were proved false

Alava

Spanish destroyer class. In 1925 the Spanish started building the first of the *Churruca* Class destroyers, whose design had been inspired by the British *Scott* Class destroyer leaders of 1918. Nine were built to the original design, of which two were sold to Argentina. All survived the civil war, and the last was disposed of in 1966.

During the late 1920s and early 1930s a further seven, the *Almirante Antequera* Class, were built to a slightly modified design. One was sunk in 1937 but was salvaged and returned to service in 1939. Two survived into the 1970s.

In 1936 the Spanish navy decided to build two more, *Alava* and *Liniers*, to a further modification of the basic design. The civil war intervened, and construction was delayed. Work restarted in 1939, but stopped again in 1940. Building did not finally start in earnest until 1944. Even then, Spain's poor financial position meant that they were built very slowly.

	laid down	launched	completed	completed reconstruction
Alava	12/1944	5/1947	12/1950	1/1962
Liniers	1/1945	5/1946	1/1951	9/1962

As first completed, they had an armament of 4 4.7-in, 6 37-mm and 3 20-mm guns, and 6 21-in torpedo tubes in two triple mountings.

With the advent of American military aid, they were modernized in 1961/62, being rearmed as fast antisubmarine frigates. They were given two side-launching racks for antisubmarine torpedoes, and a completely new antiaircraft gun armament. A lattice mast with a prominent search radar was also fitted. They can easily be recognized because the break in the forecastle comes most unusually between the two funnels.

Displacement: 1842 tons (standard), 2287 tons (full load) *Length:* 102.5 m (336 ft 3 in) oa *Beam:* 9.6 m (31 ft 6 in) *Draught:* 6.0 m (19 ft 8 in) *Machinery:* 2-shaft geared steam turbines, 31 500 shp=29 knots *Armament:* 3 3-in (76-mm); 3 40-mm; 2 hedgehogs; 8 DC mortars; 6 DC racks; 6 torpedo launching racks (2×3 above water, beam) *Crew:* 224

Albacore, Fairey

British naval biplane torpedo bomber. The Fairey Albacore was an example of a combat aircraft which—though in no sense a failure—was taken out of production earlier than the machine it was intended to replace. Designed to meet Specification S.41/36, it was a carrier-based torpedo bomber biplane intended to replace the same company's Swordfish. It came out larger and much heavier, despite the use of the extremely neat Taurus sleeve-valve engine, so that performance was only marginally improved and range was inferior. The three-man crews did not all welcome the change to an enclosed cabin, despite the introduction of electric heating, and the switch to a stressed-skin monocoque fuselage merely made repairs more difficult. The untapered wings had fabric covering, slats, and flaps which could serve as dive-bombing airbrakes. Another advance was a constant-speed propeller (the old 'Stringbag' having a simple fixed-pitch screw), and there was nothing wrong with the 'Applecore's' handling qualities. The 800 production machines, delivered from March 1940, all served with the Fleet Air Arm, and gained distinction (but no publicity) in countless bitter actions including Cape Matapan and the sustained night offensive by shore-based squadrons against the Afrika Korps in 1941-43.

Powerplant: One 1065-hp Bristol Taurus II *Span:* 15.2 m (50 ft 0 in) *Length:* 12.1 m (39 ft 9 in) *Gross weight:* 5715 kg (12 600 lb) *Maximum speed:* 259 km/h (161 mph)

Albany

US missile cruiser class. In the mid-1950s, two *Baltimore* Class heavy cruisers and six *Cleveland* Class light cruisers were converted by the US Navy to have antiaircraft missile launchers in place of the aft turrets. This was a temporary measure pending the introduction of purpose built missile ships, of which the first, the nuclear powered *Long Beach*, was laid down in December 1957.

It was obvious, however, that the new guided missile cruisers would not be available in sufficient numbers for many years, so three heavy cruisers were selected to be fully converted into guided missile cruisers. The three originally chosen were *Chicago* (CG II ex-CA 136) and *Fall River* (CA 131) of the *Baltimore* Class and *Albany* (CG 10 ex-CA

US Navy

USS *Albany* (left) after conversion to a missile cruiser in the late 1950s, and (right) landing a Sikorsky Hoverfly on her forward 8-in turret in her original form as an *Oregon City* Class cruiser

The Fairey Albacore was designed as a replacement for the Swordfish torpedo bomber, but was never as popular. Armed with a single fixed Vickers in the starboard wing and two Vickers K machine-guns in the cockpit, it could carry an 18-in (46-cm) torpedo under the fuselage or six 250-lb (113-kg) or four 500-lb (226-kg) bombs on wing racks

US Navy

On her conversion, *Albany* was fitted with two twin Talos, two twin Tartar and one Asroc launcher: here two Talos (fore and aft) and one Tartar (amidships) are fired in salvo

123) of the *Oregon City* Class, but *Fall River* was not suitable for conversion, so *Columbus* (CG 12 ex-CA 74), also of the *Baltimore* Class, was chosen to replace her.

	laid down	launched	commissioned as CA	commissioned as CG
Albany	3/1944	6/1945	6/1946	11/1962
Chicago	7/1943	8/1944	2/1945	5/1964
Columbus	6/1943	11/1944	6/1945	12/1962

The USN planned to convert two more *Baltimore* Class cruisers but the very high cost of conversion and the greater effectiveness of the new purpose-built ships meant that this plan was dropped.

The hull and machinery of the two classes were identical and since all their original armament and superstructure was removed, the three ships as converted were almost indistinguishable. The new superstructure is mainly aluminium to save weight and improve stability, and instead of having separate masts and funnels, they are combined into two 'macks'. These have vents set at an angle near the top to exhaust the steam away from the radar, and are covered with plastic in an attempt to reduce the changes in alignment of the mack-head radar caused by expansion and contraction from the varying heat of the exhaust steam.

Albany was converted in Boston naval yard from January 1959 to November 1962. She was fitted with a twin Talos long range surface-to-air missile launcher fore and aft with 92 missiles, and a twin Tartar medium range surface-to-air missile launcher on either side of the fore-bridge with 80 missiles. It was planned to mount 8 Polaris missile tubes amidships, and space was provided, but this requirement was cancelled in mid-1959.

She was given an extensive antisubmarine capability, including Sonar and an Asroc 8-tube launcher. Initially, no guns were fitted, but it was soon realized that the missiles were not suited for close low-level defence, nor for such tasks as stopping a small ship for examination. Therefore a single 5-in gun was added on either side of the after mack.

Albany's antiaircraft systems were extensively modified at Boston naval yard from February 1967 to August 1969. The refit gave *Albany* better radar and a faster and more reliable Talos fire control than the one with which *Chicago* shot down a MiG off North Vietnam at a range of 77 km (48 miles) in May 1972.

Chicago and *Columbus* were not modernized. The state of their aging hulls and the availability of new ships meant that they could soon be discarded. *Columbus* was decommissioned in January 1975. All conventionally powered cruisers are scheduled for replacement by the late 1970s, to be replaced by nuclear powered vessels.

Displacement: 13 700 tons (standard), 17 500 tons (full load) *Length:* 205.13 m (673 ft) *Beam:* 21.34 m (70 ft) *Draught:* 8.23 m (27 ft) *Machinery:* 4-shaft geared steam turbines, 120 000 shp=33 knots *Armament:* 2 Twin Talos launchers, 2 Twin Tartar launchers, 2 5-in (130-mm), 1 Asroc 8-tube launcher, 2 Triple torpedo launchers (beam) *Crew:* 1000

Albatros Italian air-to-air/surface-to-air missile See **Aspide**

Albatros

Albatros

Italian torpedo boat. This small vessel was rated as a submarine chaser when first completed, but in 1938 she was reclassified as a torpedo boat or small destroyer. She was designed by Colonel Leonardo Fea.

The *Albatros* was laid down in 1931 by Cantieri Navali Riuniti, Palermo, launched in May 1934 and completed the following November. Her recognition letters were 'AA'. She was torpedoed by the British submarine *Upright* on September 27, 1941.

Displacement: 339 tons (normal) *Length:* 70.5 m (231 ft 4 in) oa *Beam:* 6.0 m (19 ft 8 in) *Draught:* 1.71 m (5 ft 7 in) mean *Machinery:* 2-shaft geared turbines, 4000 shp=24.5 knots *Armament:* 2 100-mm (4-in) (2×1); 2 37-mm (1.5-in) AA (1×2); 2 13.2-mm (0.52-in) AA; 2 depth-charge throwers; 2 45-cm (17.7-in) torpedo tubes

Albatros

Russian torpedo boat class. Like the German navy, the Russian navy referred to light destroyers as torpedo boats. This class of eight units was started at some time before the outbreak of war, probably 1939/40. Five building at Nikolaiev in the Black Sea were captured incomplete when German forces overran the building yard but the three building in the Far East (*Albatros, Krechet* and *Chaika*) were allowed to continue. However it is not known in the West if the second and third units of this group were ever completed.

The *Albatros* is the only one of the class known to have been completed, in 1944. Her designed armament of 100-mm (3.9-in) guns was not available and so she received three 3.5-in (85-mm) guns, and four machine-guns in place of the intended 37-mm (1.5-in) AA weapons. Three 17.7-in (45-cm) torpedo-tubes were mounted in place of the 21-in (53-cm) tubes planned. Only two boilers were shipped and so the ship could only develop 12 300 shp, equivalent to 25 knots. She probably received full armament and other equipment after the end of the Second World War.

Displacement: 920 tons (normal) *Length:* 85 m (279 ft) oa *Beam:* 8.4 m (27 ft 6 in) *Draught:* 3 m

(9 ft 9 in) *Machinery:* 2-shaft geared turbines, 24 600 shp=30 knots *Armament:* (designed) 3 100-mm (3.9-in) (3×1); 6 37-mm (1.5-in) AA (6×1); 8 machine-guns; 3 53-cm (21-in) torpedo tubes (1×3); 20 mines *Crew:* 92

Albatros C

German general purpose biplane. During the First World War the German Albatros Werke, with a design team led by Dipl Ing Robert Thelen, produced some of the best and most widely used of all the Imperial Aviation Service's combat aircraft.

The first armed Albatros was the C.I two-seat general-purpose machine of 1915, built by several companies. These robust machines, powered by a 150-hp Benz III, gave excellent service. They were succeeded by even larger numbers of the C.III, which entered service in late 1915. Like the unarmed B.III, the C.III had a fish-style tail that not only improved handling but became a distinguishing feature of the Albatros single-seat scouts. The ply-covered fuselage had a roomy rear cockpit for the observer, with Schneider ring mount for the Parabellum machine-gun. Nearly all also had a fixed Spandau (almost the same 7.92-mm gun) firing ahead on the right side of the engine, and equipped with synchronization gear to enable it to fire between the propeller blades.

Though used mainly for reconnaissance and artillery spotting, the C.III could carry 100 kg (220 lb) of bombs in a compartment between the cockpits. Powered by the 160-hp Mercedes D.III engine, the C.III was built by seven manufacturers and served on all fronts until mid-1917 when it was withdrawn for use in the training role.

Span: 11.7 m (38 ft $4\frac{1}{2}$ in) *Length:* 8 m (26 ft 3 in) *Gross weight:* 1353 kg (2983 lb) *Max speed:* 141 km/h ($87\frac{1}{2}$ mph)

Albatros D

German triplane fighter. Though deadly in its day, the Fokker E monoplane was soon outclassed by the Nieuport XI and DH.2, and a more powerful machine was needed. The best of the offerings of the German aircraft industry in 1916 was the Albatros D.I, which appeared in August of that year and, after

Above: **Albatros BIII, one of the reconnaissance biplanes designed by Ernst Heinkel in 1913-14, which served with the German air force in the early part of the First World War. Powered by 100-hp Mercedes D.I or 120-hp D.II, they were among the best of their kind in 1914. In the spring of 1915 the machine-gun armed C-series were introduced, and after a variety of models the C.X (opposite) entered service. Powered by a 240-hp Mercedes D.IVa engine, this type had an increased maximum speed of 175 km/h (109 mph)**

Bayer. Hauptstaatsarchiv/Munich

The Albatros D.III of Leutnant Verner Voss, fourth on the list of Germany's First World War aces with 48 victories. ***Below:*** **A squadron of Albatros D.IIIs on the ground. Introduced in early 1917, the D.III equipped all 37 of Germany's fighter squadrons by the spring of that year, and with double the armament of contemporary Allied fighters took a heavy toll during 'Bloody April'**

very fast and troublefree development, went into action with the *Jastas* (fighter flights) the following month.

A trim, streamlined single-bay biplane, the D.I introduced the low aspect-ratio, curved 'fishtail' that became a distinguishing feature of all the Albatros scouts. It had rectangular wings of short span, a 160-hp Mercedes D.III, and could carry a pair of synchronized 7.92-mm Spandau machine-guns without suffering any loss in performance (which no previous German fighter could do).

In October 1916 the D.I was replaced in production by the D.II, with the upper wing moved lower (almost touching the windscreen) on splayed-out struts to improve pilot view. Further improvement came from moving the radiator from the sides of the nose to the interior of the upper wing. By January 1917 some 214 D.IIs at the Western Front had tipped the scales in favour of Germany.

In late 1916 Albatros designer Thelen redesigned the wings to incorporate features of the Nieuport scouts, notably to extend the outer chord of the upper planes and cut down the chord of the lower, joining them by V-struts. Allied with the 175-hp Mercedes

D.IIIa high-compression engine, the result was the best fighting scout on the Western Front, and by the spring of 1917 the D.III, as it was designated, equipped all Germany's 37 Jastas.

Some of the earliest D.IIIs were assigned to Jagdstaftel 11, the squadron commanded by the famed German ace, Baron Von Richtofen. It was Richtofen who developed the tactics of the 'flying circus'—a coordinated fighting unit—which rapidly put an end to the days of individual aerial duels.

Equipped with the D.III, the German fighter squadrons rapidly asserted their superiority over their Allied opponents in the early months of 1917, culminating in 'Bloody April'. During that month the Royal Flying Corps, in spite of Allied superiority in single-seat fighters of 358:114, suffered aircrew losses totalling over 40%, and the average life expectancy of a British airman on the Western Front dropped to 23 days.

There were many minor variations, and Oeffag (Austrian-built) models had Austro-

***Above:* The Albatros D.Va flown by Leutnant Von Hippel of Jasta 5. An improved version of the D.III with more powerful 180-hp Mercedes D.IIIa engine, it did not show sufficient improvement in performance over the earlier type. *Below:* A crashed Austrian-built D.III**

Albatros W.4

Daimler engines of up to 225 hp. Nearly 500 D.IIIs were at the Western Front in late 1917, but Albatros found it impossible to keep it ahead of the Allies' Triplane, Camel, SE.5a and Spad VII.

The Albatros D.V was very similar to the D.III but had a lighter and stronger oval-section plywood fuselage, and sometimes fractionally more power. No fewer than 1512 D.V and Va models served on the Western Front, plus many others in Macedonia, Palestine and northern Italy. Like a few D.IIIs, the D.V suffered from catastrophic failure of the slim lower wings, so that prolonged dives were prohibited, but these aircraft were still used in greater numbers than any other German fighter right up to the Armistice.

(D.I) *Span:* 8.5 m (27 ft 10¾ in) *Length:* 7.4 m (24 ft 3¼ in) *Gross weight:* 896 kg (1976 lb) *Max speed:* 175 km/h (109 mph)

(D.II) *Span:* 8.5 m (27 ft 10¾ in) *Length:* 7.4 m (24 ft 3¼ in) *Gross weight:* 881 kg (1958 lb) *Max speed:* 175 km/h (109 mph)

(D.III) *Span:* 9 m (29 ft 8¼ in) *Length:* 7.3 m (24 ft 0½ in) *Gross weight:* 886 kg (1953 lb) *Max speed:* 186 km/h (116 mph)

(D.V) *Span:* 9 m (29 ft 8¼ in) *Length:* 7.3 m (24 ft 0½ in) *Gross weight:* 937 kg (2066 lb) *Max speed:* 186 km/h (116 mph)

The Albatros D.V showed several improvements over the D.III. Unfortunately, the almost eliptical fuselage section, bigger propeller, and new ailerons and rudder were not enough to make the type competitive in performance with newer Allied designs, and a weakness of the wing caused several to break up in dives. Nevertheless, the type was used in greater numbers than any other German fighter of the First World War

Albatros W.4

German reconnaissance floatplane. A development of the D.I landplane, the W.4 was the chief combat seaplane of the Imperial German Naval Air Service prior to the introduction of Hansa Brandenburg designs. Larger than the D.I, the W.4 was a two-seater powered by the 160-hp Mercedes D.III, and normally armed with one or two Parabellum 7.92-mm machine-guns in the rear cockpit.

Span: 9.49m (31 ft 2 in) *Length:* 8.50 m (27 ft 10¾ in) *Gross weight:* 1070 kg (2359 lb) *Maximum speed:* 161 km/h (100 mph)

Albatros Aero L-39

Czechoslovakian advanced trainer. In conformity with its policy of hiving off secondary tasks to its satellites in the Warsaw Pact, the Soviet Union assigned all manufacture of trainer aircraft to Czechoslovakia and Poland, the latter producing the TS-11 Iskra and the Czechs the L-29 and L-39 Albatros.

The Albatros, first flown on November 4, 1968, is the L-29's successor as the standard trainer of all Warsaw Pact countries except Poland, and it is also in production in a light attack version, mainly for export. A finely profiled tandem-seater, with the instructor in the raised rear ejection seat giving an excellent view, it is powered by the 1720-kg (3792-lb) thrust Walter Titan, the Czech-built Ivchenko AI-25TL turbofan also used in the Yak-40 STOL trijet. The name Albatros is used only for L-39 aircraft in service with the Czech air force, to which deliveries began in April 1974.

The basic aircraft is equipped with a camera gun and electrically controlled gunsight, but has no internal gun. Gun pods, bombs, rockets and missiles can be carried on two or four wing pylons, and Aero Vodochody Narodni Podnik (which perpetuates the name of the old Aero company) is developing the L-39Z primarily for the light attack role.

Span: 9.4 m (31 ft 0½ in) *Length:* 12.25 m (40 ft 5 in) *Gross weight:* 4535 kg (9998 lb) *Max speed:* 750 km/h (466 mph)

Albatross

Australian seaplane carrier. In the period after the First World War, when the aircraft carrier was still unproven, many experts regarded the seaplane carrier as a more useful type, particularly for the Pacific. In 1926 the Royal Australian Navy laid down a seaplane carrier to its own design, and the ship which resulted was more a demonstration of the somewhat limited facilities of Cockatoo dockyard than of Australian sophistication.

HMAS *Albatross* was launched on February 23, 1928 and completed late the same year. She was an unusually ugly ship, with a high forecastle housing the hangar for nine Fairey floatplanes. The aircraft were launched from a catapult on the roof of the hangar and recovered by three cranes stepped forward of the bridge.

In 1938 she was transferred to the Royal Navy in part payment for two new cruisers. At first she was regarded as a white elephant but during the early part of the Second World War she was re-equipped with six Walrus amphibians and performed useful work on the convoy routes in the South Atlantic. Her normal base was Freetown and her aircraft carried out antisubmarine patrols.

The introduction of escort carriers in 1942-43 made the *Albatross* redundant, and in 1943 she paid off for conversion to a depot ship for escorts and ocean minesweepers. This involved the removal of the two 4.7-in AA guns from the forecastle and conversion of the hangar to a workshop. She was sold for mercantile use in August 1946 and renamed *Pride of Torquay* but was later renamed *Hellenic Prince*. She was scrapped at Hong Kong in August 1964.

Displacement: 4800 tons (standard), 6000 tons (full load) *Length:* 135.2 m (443 ft 9 in) oa *Beam:* 18.5 m (60 ft 9 in) over 'bulges' *Draught:* 4.9 m (16 ft 3 in) full load *Machinery:* 2-shaft singled reduction geared turbines, 12 000 shp=21 knots *Armament:* 4 4.7-in (120-mm) AA (4×1); 2 2-pdr pom-poms (2×1); 6 20-mm AA (6×1) added 1941 *Aircraft:* 9 (1 catapult) *Crew:* 450

Albatros D.Va. Note the auxiliary strut from the leading interplane strut to the leading edge

Albatross HU-16 Grumman

US general purpose amphibian. Last and greatest of Grumman's long series of twin-engined amphibians, the Albatross was first ordered as the XJR2F-1 utility transport for the US Navy and the prototype flew on October 24, 1947. By far the biggest original user was the US Air Force, which bought 305 SA-16A sea/air rescue models, most of which were converted to SA-16B form with extended wings and tail for operation at greater weight.

Powered by two 1425-hp Wright Cyclone R-1820-76A engines, the Albatross in US service was later designated HU-16, with suffix letters for many navy, air force and coast guard versions. Several nations used the Albatross, including Argentina, Brazil, Canada, Chile, Greece, Indonesia, Italy, Japan, Norway, Philippines, Portugal and Spain. Many still in service are of the anti-submarine type first flown in 1961, with nose radome, tail MAD boom, underwing search-light, ECM installations and provision for carrying antisubmarine depth charges, torpedoes or other stores. Full all-weather electronics and anti-icing equipment are standard, and the ability to operate from airfields or marine bases is becoming increasingly rare. The Albatross is the only military amphibian in wide service outside Japan and the Soviet Union.

Span: 29.5 m (96 ft 8 in) *Length:* 19 m (62 ft 10 in) *Gross weight:* 17 010 kg (37 500 lb) *Max speed:* 380 km/h (236 mph)

Flight International

The Armstrong Whitworth Albemarle, 600 of which were produced between 1941 and 1943

Albemarle, Armstrong-Whitworth

British bomber. Designed to Air Ministry bomber specification B 18/38, the Armstrong Whitworth AW.41 Albemarle was developed under great handicaps. At an early stage it was redesigned to use little aluminium, instead having a structure of steel tubes and plate, wood and other materials. It was then subcontracted entirely outside the aircraft industry, the parts being made by furniture and light engineering firms, and being assembled at a plant built by the Hawker Siddeley Group near Gloucester called 'AW Hawksley Ltd', completely separate from any design staff. Despite this, 600 Albemarles were produced between 1941 and 1943, the AW-built prototype having flown in March 1940.

Powered by two 1590-hp Bristol Hercules XI sleeve-valve engines, the Albemarle was one of the first British aircraft with a tricycle landing gear. Pleasant to fly, it was widely used as a glider tug, although it suffered from overheating through sustained high power at low airspeed. There were many GT (general transport) and ST (special transport) versions, some equipped with four-gun dorsal turrets (a few had a two-gun belly turret) or twin manually aimed dorsal guns. Most could carry freight, paratroops or special equipment. Albemarles were prominent in the assaults on Sicily, Normandy and Arnhem, and one batch was supplied to the Soviet Union.

Span: 23.5 m (77 ft 0 in) *Length:* 18.25 m (59 ft 11 in) *Gross weight:* 16 556 kg (36 500 lb) *Max speed:* 426 km/h (265 mph)

Alberico da Barbiano

Italian light cruiser class. Although the Mediterranean can be quite stormy, ships operating there do not need to be as seaworthy or as strongly built as those in the North Atlantic or Pacific. Long range is also of lesser importance. Consequently the *Alberico da Barbiano* Class were very lightly built and armoured and had a short range, enabling them to have a very high speed and a heavy armament on a small displacement.

They were the first Italian light cruisers to be built after the First World War, and there were four in the class:

	laid down	launched	completed
Alberico da Barbiano	4/1928	8/1930	6/1931
Alberto di Giussano	3/1928	4/1930	2/1931
Bartolomeo Colleoni	6/1928	12/1930	2/1932
Giovanni delle Bande Nere	10/1928	4/1930	4/1931

They were all named after Condottieri (Italian mercenary leaders of the late Middle Ages), as were the four classes of Italian light cruiser developed from the *da Barbianos*.

Italian propaganda made great play with the high speed made by the *da Barbianos* on their trials (one made over 42 knots). However, whereas the British and the Americans ran their trials under service conditions, the Italians tended to run theirs with the ships as light as possible. The result was that the trials speeds were not achieved in service, and by June 1940 none of this class could make more than 30 knots.

Thus their main asset, high speed, had disappeared and they were easily outclassed by their heavier-built British equivalents, which were better able to withstand battle damage and could also maintain their designed speed in service.

None of the *da Barbianos* survived the war. *Bartolomeo Colleoni* was overwhelmed in July 1940 by HMAS *Sydney* and four British destroyers. Then *Alberico da Barbiano* and *Alberto di Giussano* were caught by three British and one Dutch destroyer off Cape Bon in December 1941. Laden with petrol, they did not see the destroyers approaching from the landward side, and they were torpedoed before they had fired a shot. *Giovanni delle Bande Nere* was the last to go, torpedoed by HM Submarine *Urge* in April 1942.

Displacement: 5200 tons (standard), 7000 tons (full load) *Length:* 169.3 m (555 ft 5 in) oa *Beam:* 15.5 m (50 ft 10 in) *Draught:* 4.9 m (16 ft 8 in) *Machinery:* 2-shaft geared steam turbines, 95 000 shp =37 knots *Protection:* 25 mm belt; 20 mm deck *Armament:* 8 6-in (152-mm); 6 3.9-in (90-mm); 8 37-mm; 8 13.2-mm; 4 21-in (53-cm) torpedo tubes (2×2 above water, beam) *Aircraft:* 2 *Crew:* 521

Albion British aircraft carrier See ***Hermes***

The only military amphibian still in widespread service outside the USSR and Japan, the Grumman HU-16 Albatross has been built in coast guard (above) navy (below) and air force versions for over a dozen nations

Grumman Aerospace

The biggest customer for the Albatross was the US Air Force, which bought 305 SA-16A (Navy designation HU-16) air-sea rescue versions of the aircraft. The majority of these were subsequently converted to SA-16Bs (shown) with increased wingspan and taller tail

Albion

British aircraft/commando carrier. HMS *Albion* was the second unit of the *Hermes* Class of aircraft carriers laid down in 1944-45. She was launched on May 6, 1947 but as the design of the class was under review the hull was immediately laid up. Machinery had been installed but she was minus armament and equipment.

Work proceeded slowly, and the ship was finally completed by Swan, Hunter and Wigham Richardson at Wallsend-on-Tyne on May 26, 1954. On October 27 the same year she became flagship of the Aircraft Carrier Squadron in the Mediterranean, but transferred to the Home Fleet in March 1955. After a refit in May-September 1956 she left hurriedly for the Mediterranean to take part in the Suez operations.

Air strikes against Egyptian targets were begun by *Albion*'s Sea Hawk and Sea Venom aircraft at dawn on November 1, and included an attack on Al Naza field, six miles from Cairo and over 209 km (130 miles) from the Fleet. Numerous sorties were flown during the next two days, and the carrier's aircraft provided cover for the paratroops' attack on November 5. Her Skyraiders and helicopters landed medical supplies and evacuated wounded from Gamil Airport near Port Said. Before returning to Malta on November 29 the *Albion*'s air group had flown over 2000 sorties in two months.

After service in the Home Fleet, Far East and Mediterranean the *Albion* went into Portsmouth dockyard for conversion to a commando carrier, which took a year. She recommissioned in her new role in August 1962, and only a month after she had joined the Far East Fleet in November 1963 her Royal Marine Commandos had to quell a small and not too significant revolt in the Sultanate of Brunei.

She returned to England in April 1964 for a refit and so missed the start of the Indonesian confrontation, but she was sent out to the Far East as soon as her refit ended, and she was to remain there until the end of 1967. After a period in the Mediterranean in 1970 the *Albion* began her last Far East commission in March 1971, one which ended with the withdrawal from Singapore.

The last leg of *Albion*'s final commission began in April 1972 for she was now getting too old to warrant further modernization. When she paid off for the last time at Portsmouth on March 2, 1973 she had been in service almost continuously for 18 years. She was sold to a firm in July 1973 for conversion to a heavy lift ship to operate in the North Sea oilfields, but in November 1973 her new owners announced that the conversion would not be carried out, and the carrier was ultimately scrapped.

Displacement: 22000 tons (standard), 27000 tons (full load), 23300 tons (standard as commando carrier) *Length:* 224.8 m (737 ft 9 in) *Beam:* 27.4 m (90 ft), extreme width 37.5 m (123 ft) *Draught:* 8.2 m (27 ft) *Machinery:* 2-shaft geared turbines, 78000 shp=28 knots *Armament:* (As built) 32 40-mm (1.5-in) Bofors AA (2×6, 8×2, 4×1); (After refit with angled deck) 20 40-mm AA (1×6, 5×2, 4×1); (As Commando Ship) 8 40-mm AA (4×2) *Crew:* 1390 (originally) increased to 1035+733 commandos

Several of the *Albrighton*, or 'Hunt Type III' destroyers were transferred to the Greek navy in 1942: the *Adrias* (below), formerly HMS *Border*, was mined in October 1943 and written off

Albrighton

British destroyer escort class. The *Albrighton* Class were laid down under the 1940 Programme and launched between 1941 and 1943. They were the third group of the 'Hunt' Class escort destroyers and are usually referred to as the 'Hunt' Class Type III. The Hunts were designed as general antisubmarine and antiaircraft escorts for convoys and were originally designated Fast Escort Vessels. They were very well armed for their size, the displacement having been restricted for cheapness and speed of construction, but as they were considered escort vessels the first two groups had no torpedo armament.

In the *Albrighton* Class, however, it was decided to fit a bank of twin torpedo tubes so that they could be employed in offensive operations against enemy shipping. To compensate for the additional weight it was necessary to reduce the number of twin 4-in mountings from three, as in Type II, to two. Most of the 'Hunt' Class were fitted with stabilizers but these were not popular and were omitted from several of the ships of Type III, the space being used for stowage.

On completion several units were manned by Allied navies but remained under Admiralty control. The *Bolebroke, Border, Modbury* and *Hatherleigh* were transferred to Greece, being renamed *Pindos, Adrias, Miaoulis* and *Kanaris* respectively; the *Haldon* became the Free French *La Combattante* and the *Eskdale* and *Glaisdale* were transferred to the Norwegians.

The *Albrighton* Class served mainly in the English Channel and Mediterranean where they proved most effective in convoy escort, antisubmarine patrols and shipping strikes. Their losses were heavy; out of the 28 ships of the class 13 were lost, as compared with 11 losses out of 45 ships in the first two groups of the 'Hunt' Class.

Seven were lost in the Mediterranean. On June 15, 1942 the *Airedale* was bombed and heavily damaged by German aircraft while escorting the Malta convoy 'Vigorous' and had to be sunk by other ships of the escort. Six months later on December 11 the *Blean* was hit by two torpedoes from the *U 443* off Oran. The first blew off her stern, and the second exploded abreast her engine room and she quickly capsized and sank. The *Derwent* was torpedoed by German aircraft while in Tripoli harbour on March 19, 1943 and was written off as a constructive total loss. The Greek *Adrias* (ex-*Border*) and the *Rockwood* were also declared constructive total losses. The former vessel had her stern blown off by a mine near Kalymnos on October 22, 1943 and the latter was severely damaged by an Hs293 glider bomb in the Aegean Sea on November 11, 1943. All three of these vessels were sold for scrap in 1945/46. The *Holcombe* was torpedoed and sunk by *U 593* in the western Mediterranean, off Bougie, on December 12, 1943 and the *Aldenham* was sunk by a mine in the Adriatic on December 14, 1944.

The remaining six casualties were lost in home waters. Two were torpedoed in the English Channel by German E-boats, the *Penylan* on December 3, 1942 and the *Eskdale* on April 14, 1943. The *Limbourne* was torpedoed by the German torpedo boat *T 22* off the coast of France on October 23, 1943 and, being heavily damaged, had to be sunk by the British. Two more were constructive total losses; the *Goathland*, mined off Normandy on July 24, 1944, and the *Wensleydale*, damaged in a collision in November of the same year. Both were sold for scrap in 1945. The last of the class to be lost was the Free French *La Combattante* (ex-*Haldon*) which was mined and sunk off the Humber during the night of February 22/23, 1945.

After the war several of the class served in foreign navies. Greece retained those she had manned during the war; they were officially returned to the Admiralty in 1959 and were scrapped in Greece. Greece also purchased the *Tanatside* and *Catterick* in 1946 and renamed them *Adrias* (after the vessel lost

HMS *Goathland* (above) was mined off Normandy in July 1944, and *Aldenham* was mined in the northern Adriatic the following December. The 'Hunt' Type destroyers were well suited to escort work, with their 4-in high-angle AA guns and handiness against submarines

during the war) and *Hastings* respectively; both were scrapped in 1963. The *Glaisdale* was sold to Norway in 1946 and renamed *Narvik*; she was sold for scrap in 1961. The *Albrighton* and *Eggesford* were sold to the Federal German navy in 1959 and renamed *Raule* and *Brommy* respectively. The remaining ships of the class were sold for scrap between 1953 and 1961. See also *Atherstone, Blankney, Brissenden.*

Albrighton, Airedale—built by J. Brown.
Aldenham, Belvoir, Eskdale, Glaisdale—built by Cammell Laird.
Blean—built by Hawthorn Leslie.
Bleasdale, Catterick, Derwent, Hatherleigh, Haydon, Penylan, Rockwood—built by Vickers-Armstrong.
Bolebroke, Border, Melbreak, Modbury—built by Swan Hunter.
Easton, Eggesford, Stevenstone, Talybont—built by White.
Goathland, Haldon—built by Fairfield.
Holcombe, Limbourne—built by Stephen.
Tanatside, Wensleydale—built by Yarrow.

Displacement: 1050 tons (standard) *Length:* 85.34 m (280 ft) oa *Beam:* 9.6 m (31 ft 6 in) *Draught:* 2.36 m (7 ft 9 in) *Machinery:* 2-shaft turbines, 19 000 shp=27 knots *Armament:* 4 4-in (100-mm) (2×2); 4 2-pdr pom-poms (1×4); 2 20-mm (2×1); 2 21-in (53-cm) torpedo tubes (1×2); 4 depth-charge throwers (110 depth-charges) *Crew:* 170

Alcione Z.1007 CRDA Cant

Italian bomber. Next to the SM.79 Sparviero, the Alcione (Kingfisher) was the most important Italian bomber of the Second World War. Designed by the CRDA Cant, it was the company's first landplane, and stemmed directly from the Z.506 Airone. The prototype, flown in late 1937, had three Asso liquid-cooled engines in annular cowls, but all production machines had radials, usually the 1000-hp Piaggio P.XIbis RC40.

Of all-wooden construction, the Alcione was an efficient aircraft with a long bomb bay under the wing able to carry up to 2000 kg (4410 lb). There were manual gun positions at front and rear of this bay (often the front position was used solely for bomb-aiming) and most Alciones had a dorsal turret and two beam guns. The guns were at first all 7.7 mm, but by 1940 the dorsal and ventral guns were 12.7 mm. When Italy entered the war in 1940 there were 87 in service, and the type was being built by Cant, Piaggio and IMAM Meridionali. In 1941 the Z.1007bis appeared, with fuselage 244 cm (96 in) longer, span 83 cm (33 in) greater and stronger landing gear. Many of these had twin fins.

The Alcione operated from Tunisia to the Soviet Union, often carrying two 450-mm

Single-fin version of the Alcione, one of Italy's most important Second World War bombers

(1000-lb) torpedoes or in the strategic reconnaissance role. A few Z.1007ter, with 1175-hp Piaggio P.XIX engines, entered service in 1942. Cant tried to improve performance by fitting the 1500-hp Piaggio P.XIX engine, and at the very end of Italian participation in the war produced the excellent Z.1018 Leone (Lion). As was often the case with Italian programmes of the Second World War, the Alcione was an excellent aircraft but it could not be built in adequate numbers and the courage of its crews was frustrated by circumstances.

Span: 24.8 m (81 ft $4\frac{1}{2}$ in) *Length:* 18.3 m (60 ft $2\frac{1}{2}$ in); bis Ser. XII, 20.74 m (68 ft $2\frac{1}{2}$ in) *Gross weight:* 13 621 kg (30 029 lb) *Max speed:* 455 km/h (283 mph)

ALCM

USAF Air-Launched Cruise Missile. Developed by the US Air Force, the ALCM gives a new flexibility to bombers of Strategic Air Command. When launched in large numbers from B-52 and B-1 aircraft, each of the nuclear-armed cruise missiles would have to be countered, making defence against them difficult and costly.

According to Boeing Aerospace Company, the prime contractor, the ALCMs are intended to saturate defences and improve the ability of manned aircraft to penetrate to major targets. Although subsonic in operation their small radar image and low-altitude flight capability will enhance their effectiveness.

The ALCM—virtually a modern version of a flying bomb—is a triumph of ingenuity. Only about 4.3 m (14 ft) long, it is pre-packed into a cigar-like configuration for compact storage aboard the aircraft. After it has been released several actions take place in sequence. The air intake for a small turbofan engine pops up at the back of the fuselage, the tail surfaces spring into position, the engine ignites and the wings swing open like knife blades. Full engine thrust is rapidly achieved as a function of altitude and the missile is ready to respond to commands from its automatic guidance system.

A Boeing B-52 can carry 12 ALCMs on the wing and eight internally in a rotary launcher. The B-1 can accommodate 24 of them, all internally. The missiles are fully compatible with the USAF's Short Range Attack Missile (SRAM) on the same mounts, and the B-52 and B-1 could carry a mix of ALCMs and SRAMs to meet a variety of mission requirements.

A major advantage claimed for the ALCM is the ability to hedgehop below the radar screen while making diversionary sweeps to skirt known defence positions in hostile territory. Missiles employ a combination of inertial and terrain comparison guidance. They are kept on course by a miniature computer which compares pre-programmed geographical features with the geography the missile actually 'sees' during flight.

A series of air-launched powered trials began on March 5, 1976 from a B-52 above the White Sands Missile Range in New Mexico. The first flights, made without the full guidance system, confirmed correct operation of the engine, flight control systems and terrain-following equipment. They were followed by tests in which the inertial guidance system was updated in flight from checkpoints on the ground.

According to USAF Secretary Thomas Reed, apart from its strategic application the ALCM could have a decoy, reconnaissance and non-nuclear role in Europe.

Work on the ALCM complements US Navy activities in developing a sea-launched cruise missile in which the aim is to standardize guidance, propulsion and payload.

Length: 427 cm (168 in) *Diameter:* 61 cm (24 in) *Span:* 292 cm (115 in) *Approximate weight:* 862 kg (1900 lb) *Range:* 2200 km (1367 miles) approx *Speed:* Mach 0.55 cruise, Mach 0.7+ attack *Thrust:* (Williams Research Corporation turbofan engine) 300 kg (661 lb) approx. (Wing-mounted ALCMs can have drop tanks to achieve greater range)

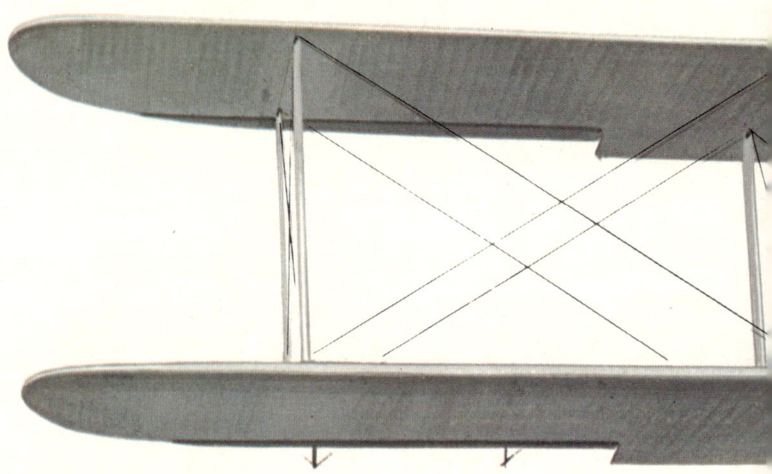

The Avro Aldershot, one of the few single-engined bombers to enter RAF service, carried a crew of three, and a bombload of only 954 kg (2000 lb)

Boeing Aerospace Company

First prototype ALCM is examined in the weapons bay of a Boeing B-52

Aldershot, Avro

British bomber. One of the least-known aircraft ever to serve with the RAF, the Avro 549 Aldershot was one of the first military machines designed after the First World War. Built to Specification 2/20 it beat the de Havilland Derby for orders, and 15, designated Aldershot III, served with specially reformed 99 Squadron at Bircham Newton between 1924 and 1926.

Powered by a 650-hp Rolls-Royce Condor III driving a large two-blade propeller, the Aldershot was large and impressive and carried a bomb load of 907 kg (2000 lb) stowed nose-up in the double-deck fuselage, a crew of three and Lewis machine-guns in the rear cockpit and aft-firing ventral position. The Aldershot was successful, and was especially useful in giving the RAF much-needed experience of navigation and bombing by night, but in 1926 the Air Staff came to the view that heavy bombers should have two engines. No 99 received Hyderabads, and Aldershots thereafter were used for research, including flight tests of the big Napier Cub and Beardmore Typhoon engines. The Andover was an ambulance version.

Span: 20.73 m (68 ft) *Length:* 13.72 m (45 ft) *Gross weight:* 4967 kg (10950 lb) *Maximum speed:* 177 km/h (110 mph).

The Boeing ALCM (Air-Launched Cruise Missile), designed to be launched in large numbers from B-52 and B-1 bombers, saturating enemy defences and increasing the chances of manned bombers reaching their targets

Alessandro Poerio

Italian scout class. Although officially designated *esploratori*, or scouts, these ships were the first of a series of large and well-armed destroyers which had an important influence on other navies during the 1920s and 1930s. They were re-rated as destroyers in 1921.

Their appearance reflected their destroyer ancestry, with a raised forecastle and two funnels. The disposition of armament was

unusual, with single guns fore and aft, singles sided in the waist and torpedo tubes staggered on either beam aft. After the First World War they were rearmed with five 4-in/45 cal guns in place of the 4-in/35 cal guns originally fitted. The disposition was changed to two guns on the forecastle, antiaircraft guns at the break of the forecastle and three guns aft.

All three ships were built by Ansaldo at Genoa, being laid down in 1913, launched in 1914 and completed in 1915. The *Cesare Rossol* was sunk by a mine off the Istrian coast in the northern Adriatic on November 16, 1918. The *Guglielmo Pepe* and *Alessandro Poerio* were transferred to Nationalist Spain in June 1938 and renamed *Teruel* and *Huesca*. The former was scrapped in 1947 and the latter in 1949.

Displacement: 896 tons (normal) 1012 tons (full load) *Length:* 83.1 m (272 ft $6\frac{1}{2}$ in) oa *Beam:* 8 m (26 ft 3 in) *Draught:* 2.83 m (9 ft $3\frac{1}{2}$ in) mean *Machinery:* 2-shaft turbines, 20000 shp=$31\frac{1}{2}$ knots *Armament:* (As built) 6 4-in (100-mm) (6×1); 2 40-mm (1.5-in) AA; 4 45-cm (17.7-in) torpedo tubes (As rearmed) 5 4-in; 2 40-mm AA; 4 45-cm torpedo tubes; minelaying equipment *Crew:* 137

Alfa

Spanish machine-guns (Ametrallador Modelo 44 and 55). The Alfa is an elderly design, having first started, as the model number suggests, in 1944. It was first chambered for the German 7.92-mm round but the 1955 model was built for 7.62-mm NATO and the earlier versions were changed during the following years. The gun was the standard issue for the Spanish armed forces until a few years ago when it was replaced by the German MG 1 which has now been bought in quite large numbers. The Alfa is now thought to be relegated to use by reserve forces and the gendarmerie.

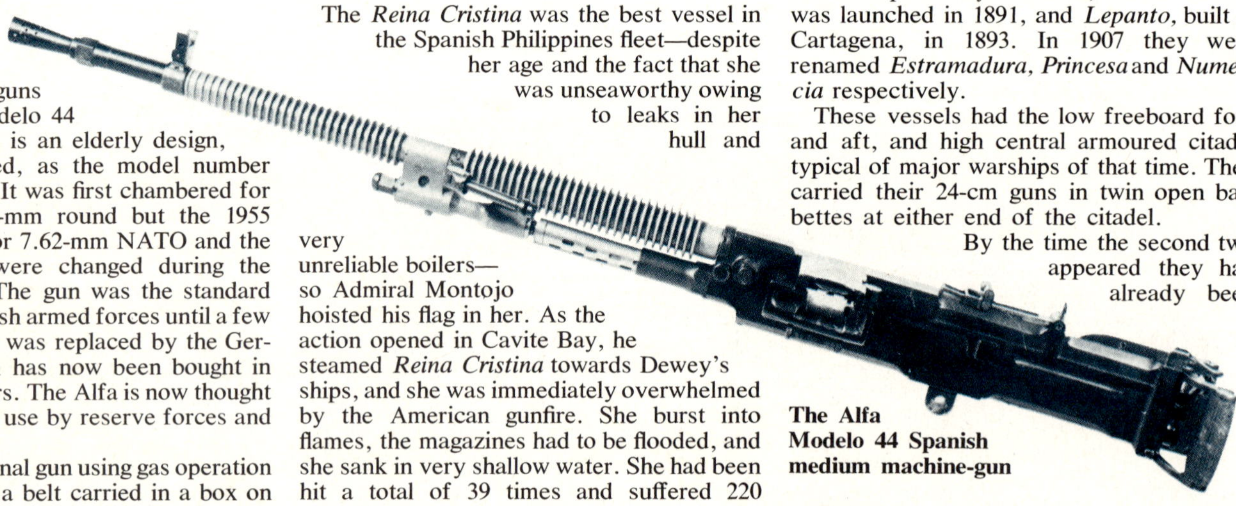

The Alfa Modelo 44 Spanish medium machine-gun

It is a conventional gun using gas operation and feeding from a belt carried in a box on the left side of the body. It is heavier than most of its contemporaries and was generally seen mounted on a tripod. However, it can be used from a bipod in the light role, but is more than a little clumsy and awkward. Most of the weight stems from the fact that the gun is made by the older techniques of machining from the solid and using generous dimensions. Replacement parts would be expensive and difficult to make today, which is probably the reason for its being taken out of service.

The Alfa was an undistinguished design which never attracted much sales interest, though some were sold to North African states.

(Modelo 5) *Weight unloaded:* 12.92 kg (28 lb 8 oz) *Length:* 1447 mm (57 in) *Operation:* gas *Calibre:* 7.62-mm NATO *Magazine:* 100 rounds belt feed *Rate of fire:* 800 rpm *Muzzle velocity:* 761/m/sec (2500 ft/sec)

Alfonso XII

Spanish cruiser class. *Alfonso XII* was one of three very similar, though not identical, cruisers built by the Spanish navy for service in Spain's overseas possessions. The first to be built was the *Reina Cristina* of 3520 tons, which was laid down in August 1881 and completed in May 1886. The second, the *Alfonso XII* of 3900 tons, was completed in August 1887, and the third, the *Reina Mercedes* of 3090 tons, in September 1887.

They all carried the same armament, and were all rigged as three-masted barques, carrying and using their sails on long passages. They had no armour protection, and were intended more for 'showing the flag' than for fighting. Indeed, *Alfonso XII* was chosen to represent Spain at the 400th anniversary celebration of the discovery of America held at Genoa in 1892.

From 1891, *Reina Cristina* was based in the Philippines, but the other two stayed in home waters until the need to reinforce the overseas squadrons to meet the threat from America became apparent in the late 1890s. *Reina Mercedes* bombarded the fortifications at Melilla as part of the Riff wars in 1893—the only action any of the three saw before the Spanish-American war.

For the next few years *Alfonso XII* and *Reina Mercedes* formed the Instruction Squadron based at Bilbao. Then all three were involved in the Spanish-American war of 1898.

The *Reina Cristina* was the best vessel in the Spanish Philippines fleet—despite her age and the fact that she was unseaworthy owing to leaks in her hull and very unreliable boilers—so Admiral Montojo hoisted his flag in her. As the action opened in Cavite Bay, he steamed *Reina Cristina* towards Dewey's ships, and she was immediately overwhelmed by the American gunfire. She burst into flames, the magazines had to be flooded, and she sank in very shallow water. She had been hit a total of 39 times and suffered 220 casualties.

Both *Reina Mercedes* and *Alfonso XII* were sent to Cuba, where both were used as depot ships, since neither was able to steam. *Reina Mercedes* was stationed at Santiago, where two of her 160-mm guns were dismounted and used in the shore batteries, as they were the only modern guns available. During the intermittent American bombardment of the Spanish fleet, the *Reina Mercedes* was hit several times and set on fire twice. Then on July 4, 1898 she was used to block the harbour entrance to prevent the American fleet steaming in, but because of the American bombardment she was not placed properly and she failed to block the channel. She was raised by the Americans after the war and after being repaired was incorporated in the USN as a training ship. She was not finally disposed of by them until November 1957.

Alfonso XII was at Havana in 1898, and was damaged by the explosion of the USS *Maine*. Her guns were dismounted and used at the Chorrera and la Reina shore batteries. At the end of the war she was used to repatriate servicemen.

It was these three ships' and their crews' misfortune that they were forced to fight in circumstances that they were not designed to combat.

Displacement: 3900 tons (normal) *Length:* 84 m (275 ft 7 in) oa *Beam:* 13 m (42 ft 8 in) *Draught:* 6.7 m (22 ft) *Machinery:* 1-shaft triple-expansion steam, 4100 ihp=15 knots *Armament:* 6 160-mm (6.3-in); 3 57-mm (2.24-in); 2 42-mm (1.65-in); 6 37-mm (1.46-in); 5 torpedo tubes (5×1) *Crew:* 380

Alfonso XIII

Spanish cruiser class. *Alfonso XIII* was one of three armoured cruisers built for the Spanish navy in the late 1880s. They were designed by Sir Nathaniel Barnaby just after he had quitted the post of Director of Naval Construction to the British Admiralty. The design was based on that of the British armoured cruiser *Australia*, launched in 1886.

The first of the class, *Reina Regente*, was built by J & G Thompson of Clydebank in Scotland, and was launched on February 24, 1887. She cost 6 million pesetas.

The other two members of the class were built in Spain. *Alfonso XIII*, built at Ferrol, was launched in 1891, and *Lepanto*, built at Cartagena, in 1893. In 1907 they were renamed *Estramadura*, *Princesa* and *Numencia* respectively.

These vessels had the low freeboard fore and aft, and high central armoured citadel typical of major warships of that time. They carried their 24-cm guns in twin open barbettes at either end of the citadel.

By the time the second two appeared they had already been completely outclassed by more modern British cruiser designs, which had a high forecastle and poop to improve seaworthiness, and enclosed barbettes for the main armament.

All three of these ships were extremely unsuccessful, being plagued by unreliable machinery throughout their short existence, as well as suffering from poor workmanship. *Reina Regente* was designed to make 20 knots, but only actually made 14 for most of her service career, and all three spent most of their time in port. They were all discarded by 1911.

Displacement: 4664 tons (normal) *Length:* 97.30 m (319 ft $3\frac{3}{4}$ in) oa *Beam:* 15.43 m (50 ft $7\frac{1}{2}$ in) *Draught:* 5.90 m (19 ft 4 in) *Machinery:* 2-shaft triple-expansion steam reciprocating, 11598 ihp=20 knots *Protection:* 82 mm ($3\frac{1}{4}$ in) deck *Armament:* 4 24-cm (9.45-in) (2×2); 6 12-cm (4.7-in); 6 57-mm (2.24-in); 1 42-mm (1.65-in); 5 torpedo tubes (5×1) *Crew:* 420

Alfonso XIII

Spanish battleship. After the disaster of the Spanish-American war, Spain gradually

Breguet 1050 Alizé antisubmarine search/strike aircraft with its radar scanner lowered. The Alizé was developed from the abandoned Vultur carrier-based attack aircraft

rebuilt her fleet. Three battleships were projected under the Naval Law of January 7, 1908. These were to be called *España*, *Alfonso XIII* and *Jaime I*, and were designed by a British Consortium of Armstrongs, John Browns and Vickers. A new shipyard, the Sociedad Español de Construcción Naval at Ferrol, was built by the consortium to construct the battleships, and the supervision and some of the skilled labour was also provided by the English shipyards.

Alfonso XIII was the second of the three to be built. Laid down on February 23, 1910 and launched on May 7, 1913, she was completed on August 16, 1915.

She took part in the bombardment of Moroccan towns during the Riff revolt in the early 1920s. In 1931 she was renamed *España* because the original *España* had run aground near Cape Tres Forcas, Morocco, on August 26, 1923, and had become a total loss.

Her next action came in the Spanish Civil War, when she was one of the four Spanish warships taken over by the Nationalists. She took part in several shore bombardments, including that on Republican positions near Bilbao on April 2, 1937. However, on April 30, 1937 she hit a mine near Santander, and soon sank, taking a number of her crew with her.

Contemporaries sneered at these, the smallest Dreadnoughts to be built, but they were a sensible design for Spain's needs. With no obvious enemies, Spain required battleships for prestige and coast defence, and for such purposes the *España* Class represented a much better investment than the large Swedish coast defence vessels, or the Brazilian *Minas Gerais* Class Dreadnought, which were their nearest equivalents.

They were unmistakable, with a flush deck and a single funnel with a twin 12-in turret *en echelon* either side, the other two twin 12-in turrets being on the centre-line fore and aft.

Displacement: 15700 tons (normal) *Length:* 133.96 m (439 ft 6 in) *Beam:* 24.00 m (78 ft 9 in) *Draught:* 7.92m (26 ft) *Machinery:* 4-shaft steam turbines, 15500 shp=19.5 knots *Protection:* 202 mm (8 in) belt, 37 mm (1½ in) deck, 202 mm turrets *Armament:* 8 12-in (304-mm); 20 4-in (100-mm); 2 3-pdr; 2 Maxim machine-guns; 3 torpedo tubes (3×1 submerged) *Crew:* 854

Algérie

French heavy cruiser. The first two classes of French interwar heavy cruisers, the two *Duquesnes* and the four *Suffrens* were disappointing. Built well under the Washington Treaty limit of 10000 tons, they were no match for their contemporaries abroad. Even though each of the *Suffrens* incorporated improvements over its predecessor, resulting in a doubling in the weight of armour between the first and the last of the class, they were too lightly built and protected, with the emphasis on speed not staying power.

Piecemeal improvement was not the answer; basic redesign was necessary. The result was the *Algérie*, the best of all the Washington Treaty cruisers. Ordered under the 1930 Programme, laid down in March 1931, and launched in May 1934, she was completed in 1934.

By accepting a 2-knot drop in speed, her designers were able to double the weight of armour over the last of the *Suffren* Class. She differed in appearance from the *Suffrens* too, having a single funnel, one armoured conning tower and a flush deck.

Although her designers undoubtedly overstepped the Treaty limitations, *Algérie*'s design was still inhibited by them. Even so, she still compared very favourably with the 40% larger German *Hipper* Class heavy cruisers.

Unfortunately, *Algérie* never had the chance to show how good she was. The only action she saw was as flagship of the First Squadron at the bombardment of Genoa in 1940. Although she took part in sweeps against German surface raiders, she never caught one.

After the French surrender, she only made a few brief sorties. Then from January to April 1942 she was extensively refitted. A French radar was fitted, and then modified to give a better performance. At the same time the mainmast and aircraft catapult were removed and replaced by an antiaircraft platform mounting an additional 16 37-mm and 16 13.2-mm guns. Unfortunately, when the Germans arrived at Toulon on November 11, 1942, *Algérie* was caught in harbour, and along with many of the best French warships was scuttled to avoid capture.

Although the days of cruisers were numbered when *Algérie* was designed, she was a fitting tailpiece to them, and along with the six light cruisers of *La Galissonnière* Class represented the high point of French interwar warship design.

Displacement: 10000 tons (standard), 13900 tons (full load) *Length:* 185.7 m (610 ft 9 in) oa *Beam:* 20 m (65 ft 8 in) *Draught:* 7.1 m (123 ft 4 in) *Machinery:* 2-shafts geared steam turbines, 84000 shp=31 knots *Protection:* 110 mm (4¼ in) belt; 76 mm (3 in) main deck *Armament:* 8 203-mm (8-in) (4×2); 12 100-mm (3.9-in) (6×2); 16 13.2-mm; 6 55-cm (21.7-in) torpedo tubes (2×3 on deck) *Aircraft* 2 *Crew:* 616

Alizé Breguet

French naval attack aircraft. Soon after the Second World War the famed French Breguet company began development of the Br.960 Vultur carrier-based attack aircraft, with Mamba turboprop boosted at the rear by a Hispano (Rolls-Royce licence-built) Nene turbojet switched in for high-speed dashes at 893 km/h. Flown in August 1951 the Vultur performed well, but the Aéronavale decided a more useful aircraft would be a comparatively slow antisubmarine search/strike machine, with no jet. The resulting Br.1050 Alizé (Tradewind) flew on October 6, 1956, and the first of 75 for the Aéronavale was delivered on March 26, 1959.

Powered solely by a 2100-hp Rolls-Royce Dart 21, the Alizé has a smaller wing than the Vultur, of different profile, with twin-wheel

An Alizé of the Indian navy, which purchased 17 in 1961. They equipped No 310 Squadron on board the aircraft carrier *Vikrant*

The Alizé carries one torpedo or three depth-charges in the fuselage, with two more depth-charges and rockets or missiles under the wings

Pre-takeoff checks for a carrier-based Alizé. The Vultur, from which the Alizé was developed, was planned to have an Armstrong-Siddeley Mamba in the nose and a French-built Rolls-Royce Nene in the rear fuselage. The intention was that it would use only the Mamba during normal cruising flight for long endurance, switching on the Nene when extra power was required for takeoff, climb and combat. When the Vultur was adapted as the ASW Alizé, speed became less essential, so the Nene was removed to make way for a retractable radar scanner and the Mamba replaced by a Rolls-Royce Dart

ECP Armées

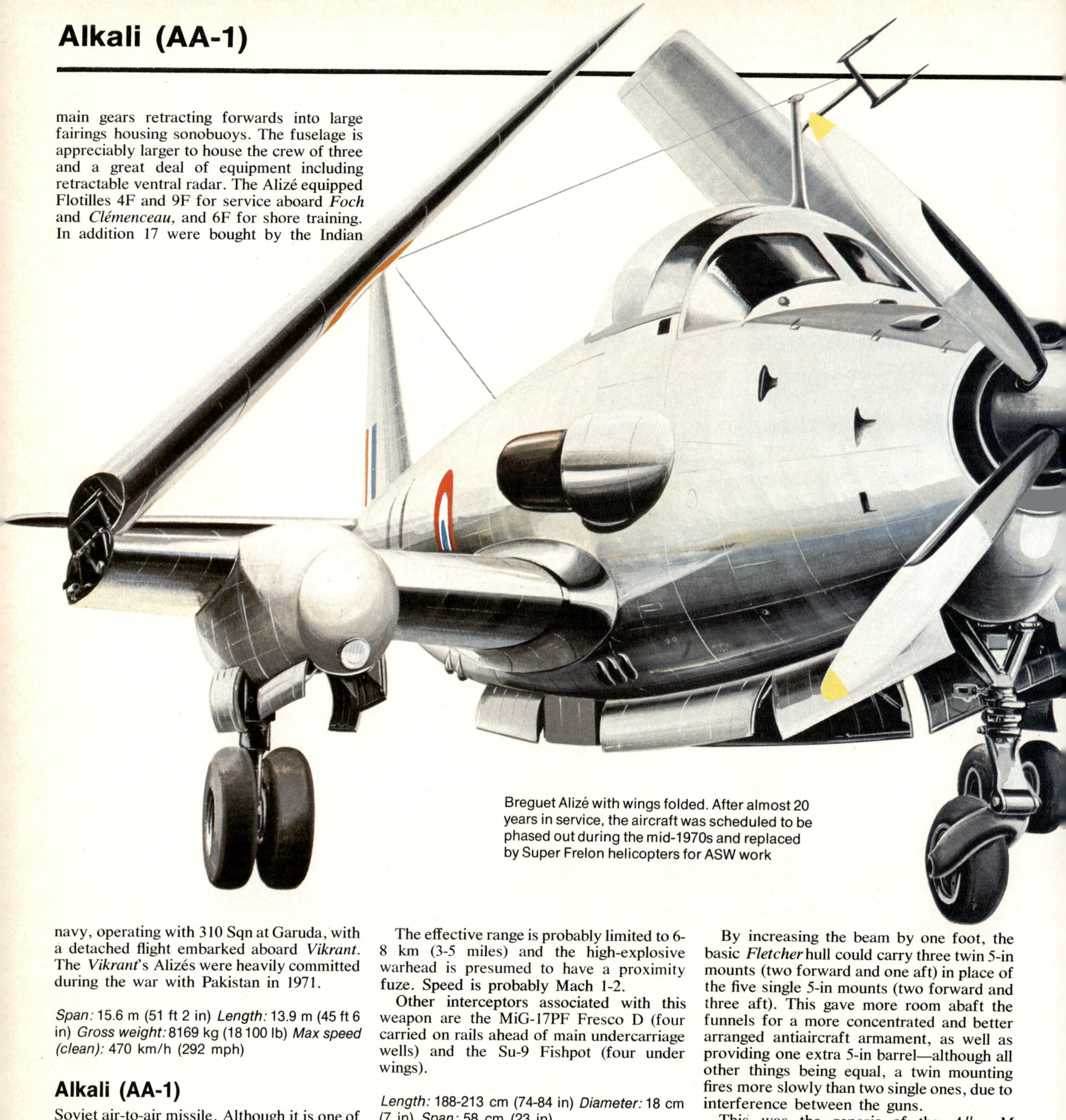

Breguet Alizé with wings folded. After almost 20 years in service, the aircraft was scheduled to be phased out during the mid-1970s and replaced by Super Frelon helicopters for ASW work

main gears retracting forwards into large fairings housing sonobuoys. The fuselage is appreciably larger to house the crew of three and a great deal of equipment including retractable ventral radar. The Alizé equipped Flotilles 4F and 9F for service aboard *Foch* and *Clémenceau,* and 6F for shore training. In addition 17 were bought by the Indian navy, operating with 310 Sqn at Garuda, with a detached flight embarked aboard *Vikrant*. The *Vikrant*'s Alizés were heavily committed during the war with Pakistan in 1971.

Span: 15.6 m (51 ft 2 in) *Length:* 13.9 m (45 ft 6 in) *Gross weight:* 8169 kg (18 100 lb) *Max speed (clean):* 470 km/h (292 mph)

Alkali (AA-1)

Soviet air-to-air missile. Although it is one of the oldest Soviet air-to-air missiles, few details of this weapon are available beyond those facts which can be deduced from photographs. Four missiles known by the NATO code Alkali were carried by the MiG-19PM Farmer all-weather interceptor on underwing pylons.

As it is a first generation weapon some reports suggest that the missile employs some form of command guidance or is beam-riding but nearly all recent accounts conclude that semi-active radar homing is employed. The missile has a cylindrical body with a tapered nosecone and a cone-shaped tail. The cruciform delta wings have small elevons which operate with four steerable foreplanes.

The effective range is probably limited to 6-8 km (3-5 miles) and the high-explosive warhead is presumed to have a proximity fuze. Speed is probably Mach 1-2.

Other interceptors associated with this weapon are the MiG-17PF Fresco D (four carried on rails ahead of main undercarriage wells) and the Su-9 Fishpot (four under wings).

Length: 188-213 cm (74-84 in) *Diameter:* 18 cm (7 in) *Span:* 58 cm (23 in)

Allen M Sumner

US destroyer class. The US *Fletcher* Class destroyers were one of the most successful class of warships to be used in the Second World War, but by 1943 it was obvious that its design could be improved.

The most obvious requirement was for better antiaircraft armament and fire-control. The obvious way to do this was to fit the very successful twin 5-in/38 cal turret used as secondary armament in battleships and cruisers and as the main antiaircraft battery in fleet carriers, in place of the single 5-in/38 cal mounting used in the *Fletchers*.

By increasing the beam by one foot, the basic *Fletcher* hull could carry three twin 5-in mounts (two forward and one aft) in place of the five single 5-in mounts (two forward and three aft). This gave more room abaft the funnels for a more concentrated and better arranged antiaircraft armament, as well as providing one extra 5-in barrel—although all other things being equal, a twin mounting fires more slowly than two single ones, due to interference between the guns.

This was the genesis of the *Allen M Sumner* Class, 70 of which were built between 1943 and 1945. They were succeeded on the stocks from 1945 to 1947 by the *Gearing* Class, which were basically the same as the *Allen M Sumners*, except that a 14-ft hull extension was inserted in the mid-section between the two funnels to accommodate extra fuel—the one major fault with the *Allen M Sumners* had been their relatively short range for Pacific warfare. The additional hull section also gave the *Gearings* a greater measure of stability.

The *Allen M Sumners* were all built at six yards, the Bethlehem Steel Corporation Staten Island yard, San Francisco yard and San Pedro yard, the Federal Ship Building and

Allen M Sumner in the South China Sea, June 1967. Seventy destroyers of this class were built in 1943-45, and 40 survived in US service until the 1970s. Of these 31, including the name ship, had been given the FRAM (Fleet rehabilitation and modernization) II conversion, which included improved sonar and electronics, ASW torpedoes and new funnel caps

US Navy

Dry Dock Company yard, the Bath Iron Works Corporation yard, and the Todd Pacific Shipyard. As built the class was divided into two groups, 58 being completed as ordinary destroyers, and 12 as light minelayers—identical with the other 58 except that they had provision to carry 80 mines.

Destroyers: *English* DD696, *Haynsworth* DD700, *John W Weeks* DD701, *Hank* DD702, *Compton* DD705, *Gainard* DD706, *Soley* DD707, *Harlan R Dickson* DD708, *Barton* DD722, *Maddox* DD731, *Hyman* DD732, *Purdy* DD734, *Brush* DD745, *Samuel N Moore* DD747, *Harry E Hubbard* DD748, *John R Pierce* DD753, *Beatty* DD756, *Henley* DD762, *Willard Keith* DD775, *Bristol* DD857, *Hugh W Hadley* DD774, *Cooper* DD695, *Meredith* DD726, *Mannert L Abele* DD733, *Drexler* DD741, *Allen M Sumner* DD692, *Moale* DD693, *Ingraham* DD694, *Charles S Sperry* DD697, *Ault* DD698, *Waldron* DD699, *Wallace L Lind* DD703, *Borie* DD704, *Hugh Purvis* DD709, *Walke* DD723, *Laffey* DD724, *O'Brien* DD725, *De Haven* DD727, *Mansfield* DD728, *Lyman K Swenson* DD729, *Collett* DD730, *Blue* DD744, *Taussig* DD746, *Alfred A Cunningham* DD752, *Frank E Evans* DD754, *John A Bole* DD755, *Putnam* DD757, *Strong* DD758, *Lofberg* DD759, *John W Thomason* DD760, *Buck* DD761, *Lowry* DD770, *James C Owen* DD776, *Zellars* DD777, *Massey* DD778, *Douglas H Fox* DD779, *Stormes* DD780, *Robert K Huntington* DD781.

Light Minelayers: *Adams* DM27, *Gwin* DM33, *Harry F Bauer* DM26, *Henry A Wiley* DM29, *Lindsey* DM32, *Robert H Smith* DM23, *Shannon* DM25, *Shea* DM30, *Thomas E Fraser* DM24, *Tolman* DM28, *J William Ditter* DM31, *Aaron Ward* DM34.

Of these, *Cooper, Meredith, Mannert L Abele* and *Drexler* were lost during the war, and *Hugh W Hadley* was so badly damaged by a kamikaze attack that she was scrapped soon after the war ended.

After the war most of the class except some of the light minelayers had their 40-mm and 20-mm guns replaced by 6 3-in and the pole mast was replaced by a tripod to carry the heavier radar. One of the two pentad 21-in torpedo tube mountings had already been removed on most to make way for a quadruple 40-mm gun mounting.

The majority were then mothballed in the reserve fleet until the early 1960s, when 29 were given the FRAM II modernization. These were: *Allen M Sumner, Moale, Ingraham, Charles S Sperry, Ault, Waldron, Wallace L Lind, Borie, Hugh Purvis, Walke, Laffey, O'Brien, De Haven, Mansfield, Lyman K Swenson, Collett, Blue, Taussig, Alfred A Cunningham, Frank E Evans, John A Bole, Putnam, Lowry, James C Owen, Zellars, Massey, Douglas H Fox, Stormes* and *Robert K Huntington.*

This conversion was intended to make the destroyers suitable for modern antisubmarine warfare. All the old antiaircraft guns were removed and the space was used to fit a hangar and landing pad for two of the unsuccessful DASH drone helicopters. Modern antisubmarine weapons and electronic equipment, including a variable depth sonar, were fitted and living spaces rehabilitated.

Country	New name	Old name	Date transferred or sold
Argentina	*Bouchard*	*Borie*	7/1972
	Segui	*Hank*	7/1972
	for spares	*Mansfield*	6/1974
Brazil	*Mato Grosso*	*Compton*	9/1972
	Alagoas	*Buck*	7/1973
	Sergipe	*James C Owen*	7/1973
	Espirito Santo	*Lowry*	10/1973
	Rio Grande De	*Strong*	10/1972
Chile	*Bordales*	*Douglas H Fox*	1/1974
	Zenteno	*Charles S Sperry*	1/1974
Colombia	*Caldas*	*Willard Keith*	7/1972
	Santander	*Waldron*	10/1973
Greece	*Miaoulis*	*Ingram*	7/1971
Iran	*Babr*	*Zellars*	1972
	Palang	*Stormes*	1972
	for spares	*Gainard*	3/1971
South Korea	*Dae Gu*	*Wallace L Lind*	12/1973
	Iu Cheon	*De Haven*	12/1973
Taiwan	*Hsiang Yang*	*Brush*	2/1976
	Heng Yang	*Samuel N Moore*	2/1970
	Hua Yang	*Bristol*	2/1970
	Yueh Yang	*Haynsworth*	5/1970
	Po Yang	*Maddox*	5/1970
	Lo Yang	*Taussig*	5/1974
	Wu Yang	*John W Thomason*	5/1974
	Huei Yang	*English*	9/1970
Turkey	*Muavenet*	*Gwin*	10/1971
Venezuela	*Carabobo*	*Beatty*	7/1972
	Falcon	*Robert K Huntington*	10/1973

The *Allen M Sumner* FRAM II conversions were not as successful as the *Gearings*, whose extra space and stability now came in useful. Despite this, the *Allen M Sumners* lasted virtually intact until the early 1970s, with none transferred to foreign navies. However, the rundown, when it came, was swift. The last of this class in US Naval service, *Laffey*, was stricken on March 1, 1975.

Twenty-nine were transferred or sold abroad, of which two were for spares: the above chart shows the purchasing countries, new names and date of transfer.

All in all, they were a successful class, but they have always been overshadowed by their famous predecessors and successors.

Displacement: 2200 tons (standard) 3320 tons (full load) *Length:* 114.76 m (376 ft 6 in) *Beam:* 12.44 m (40 ft 10 in) *Draught:* 5.79 m (19 ft) *Machinery:* 2-shaft geared steam turbines, 60 000 shp=34 knots *Armament:* 6 5-in (130-mm); 2 torpedo-launchers (triple) 2 torpedo tubes (fixed); 2 Hedgehogs *Aircraft:* 2 drone helicopters *Crew:* 274

Alligator

Russian submarine class. Four submarines were built in Russia to the designs of Simon Lake, whose ideas had been rejected by the US Navy in favour of Holland's. They exemplified Lake's ideas on submarines, with good handling on the surface and a robust

The *Allen M Sumner* in April 1953 before the FRAM II conversion. The only wartime modification to the class had been to replace the after bank of torpedo tubes with an additional quadruple 40-mm AA mounting and director. The modernization in the early 1960s was not altogether successful because of the limited space available

US Navy

hull, but proved rather unreliable in service.

The armament was heavy, comprising two internal tubes, two 'drop collars' and a twin rotating deck-tube. The *Alligator, Drakon, Kaiman* and *Krokodil* were built in 1906-11 for the Baltic Fleet and formed the 2nd Division. The *Drakon* reconnoitred Windau in September 1914, and in May 1915 she fired three torpedoes at the German cruiser *Thetis,* and so prevented her from laying a minefield. Her sister *Alligator* attacked the same ship without success a month later, and was active in the Bay of Riga in August. In October all four were sent into the Gulf of Bothnia to attack German colliers, and on October 29 the *Kaiman* captured the steamer *Stahleck.*

In August 1916 the *Krokodil* captured another steamer but at the end of November the class was taken out of service. Despite their mechanical troubles all four boats had carried out a number of patrols. In particular the *Drakon* carried out more than any other Russian submarine. Nothing is known of their subsequent fate, but it is almost certain that they fell into disrepair during the Revolution and were scuttled or scrapped subsequently. What is certain is that they were no longer in existence by 1920.

Displacement: 409/480 tons (surfaced/submerged) *Length:* 40.9 m (134 ft 5 in) *Beam:* 4.57 m (15 ft) *Draught:* 3.96 m (13 ft) *Machinery:* 2-shaft gasoline engines, 1200 ihp=$8\frac{1}{2}$ knots (surfaced), 2-shaft electric motors, approx 300 ehp=7 knots (submerged) *Armament:* 1 47-mm (1.85-in); 1 machine-gun (both added 1914-15; *Drakon* had a 37-mm AA gun); 6 45-cm (17.7-in) torpedo tubes (2 blow, 4 external) *Crew:* 20

Almirante Brown

Argentinian heavy cruiser class. In 1927 two heavy cruisers were laid down in Italy for the Armada Republica Argentina (Argentine navy). They were basically similar to the Italian *Trento* Class but were armed with only three twin 7.5-in gun turrets as against four 8-in. Another important difference was the repositioning of boilers, which gave the Argentine ships single funnels.

The *Vienticinco de Mayo* (Twenty-fifth of

May) was launched on August 11, 1929 by Orlando at Livorno, followed, on August 25, by the *Almirante Brown* from Odero-Terni's yard at Sestri Ponente. On trials both cruisers reached their contract speed of 32 knots. Their single trunked funnels gave them a passing resemblance to the British *Leander* Class, but their fire-control was typically Italian in appearance.

The names were particularly appropriate, May 25, 1810 being the date on which the revolt against Spain began, and William Brown being the Irish naval officer who had led the fleet during that war, and in the war against Brazil in 1826-28.

As completed in 1931 both ships had short funnels, but these were subsequently raised. During the Second World War both ships finally received the catapults which had been included in the original design. The catapult and crane were carried on the centreline between the funnel and mainmast, with two Grumman floatplanes. When built six twin 100-mm antiaircraft guns were mounted at forecastle deck level but these were later replaced by twin 40-mm Bofors guns.

The ships were not highly thought of, and the general opinion of them was that too much had been attempted on a small displacement. Nevertheless they lasted 30 years, and were not scrapped until 1964.

Displacement: 6800 tons (standard); 8600 tons (full load) *Length:* 170.98 m (561 ft) oa *Beam:* 17.68 m (58 ft) *Draught:* 5.63 m (16 ft 3 in) mean *Machinery:* 2-shaft geared steam turbines, 85 000 shp=32 knots *Protection:* 2¾ in (70 mm) belt; 2 in (50 mm) turrets; 1 in (25 mm) deck; 2 in (59 mm) conning tower *Armament:* 6 7.5-in (190-mm) 52 cal (3×3); 12 100-mm (3.9 in) AA (later 40-mm AA) (6×2); 6 light guns; 6 21-in (53-cm) torpedo tubes (2×3) *Crew:* 700

Almirante Clemente

Venezuelan frigate class. In the early 1950s the Americans provided what was termed 'Offshore Aid' to a number of countries, under which warships were built in Europe for various 'free world' countries, but were paid for by the USA. Most of these ships were minesweepers, but they also included the very similar *Tritan*, *Albatros* and *Pattiniwa* Class corvettes for Denmark, Italy and Indonesia respectively. These were all designed and built by Ansaldo of Italy, who also designed a rather larger fast frigate, of which six were supplied to Venezuela and two to Indonesia.

Three of the Venezuelan vessels were ordered in 1953, and the other three in 1954; the two almost identical Indonesian vessels were ordered in 1956 (see chart).

Venezuelan vessels	laid down	launched	completed
Almirante Clemente	5/1954	12/1954	1956
Almirante Jose Garcia	12/1954	10/1956	1957
Almirante Brion	12/1954	9/1955	1957
General Jose de Austria	12/1954	7/1956	1956
General Jose Trinidad Moran	5/1954	12/1954	1956
General Juan Jose Flores	5/1954	2/1955	1956
Indonesian vessels			
Iman Bondjol	1/1956	5/1956	5/1958
Surapat	1/1956	5/1956	5/1958

These ships were really light destroyers, with geared turbines rather than diesels to give a high speed. They were designed for service in tropical waters, with large airy superstructures and air-conditioning throughout the living and command spaces. Aluminium alloys were extensively used in the superstructure to reduce topweight. They were all equipped with Denny-Brown active fin stabilizers and fully automatic radar-controlled 4-in guns.

Almirante Jose Garcia, Almirante Brion and *General Jose Trinidad Moran* were refitted by the Cammell Laird/Plessy group from April 1968, the former recommissioning in February 1975 and the latter in May 1975.

These frigates are a useful cheap addition to navies that do not need sophisticated vessels. In the late 1950s and early 1960s Indonesia acquired a large number of modern Russian warships, which she found difficult to operate even before political disagreements with Russia made spares impossible to obtain. The result is that the four Italian-built frigates and corvettes of the *Iman Bondjol* and *Pattimura* Classes continued to form the backbone of the Indonesian navy after the Russian built ships had been disposed of.

Displacement: 1300 tons (standard), 1500 tons (full load) *Length:* 99.1 m (325 ft) *Beam:* 10.8 m (35 ft 5 in) *Draught:* 3.7 m (12 ft 1 in) *Machinery:* 2-shaft geared steam turbines, 24 000 shp=32 knots *Armament:* 4 4-in (100-mm); 4 40-mm (1.5-in); 8 20-mm; 3 21-in (53-cm) torpedo tubes (above water triple) *Crew:* 162

Almirante Grau/ Coronel Bolognesi

Peruvian scout cruisers. Modern warships are not renowned for their longevity but these two scout cruisers served with the Peruvian navy for the remarkable period of 51 years. For a large part of that period they were Peru's front-line ships, the *Almirante Grau* serving as fleet flagship. They were ordered from Vickers in 1905 and launched at Barrow in 1906. *Grau* was completed in the same year and the *Bolognesi* in 1907. Their lives were comparatively peaceful, for, although involved in wars with Colombia (1933) and Ecuador (1941) and taking the Allied side towards the end of the Second World War, Peru was not involved in any naval action of note.

The two ships differed in that *Grau*, being flagship, was fitted with a poop deck while *Bolognesi* was not. To keep them moderately up to date both cruisers were modified several times. During refits at Balboa in 1923-25 the boilers were converted to burn oil instead of coal, and six of the 6-pdr guns were removed. Between 1934 and 1935 the ships were reboilered by Yarrow, who had built the original installation. In the following year two of the old 3-in guns were removed and two 3-in and four 20-mm guns were added for AA defence. Also at about this time (1936) the

Almirante Clemente as refitted in 1976 with an Oto-Melara 76-mm (3-in) bow gun and Plessey AWS-1 air warning and target acquisition radar. The water-cooled automatic gun is a dual-purpose weapon whose rate of fire is reported to be as high as 150 rds/min

Oto-Melara

mainmasts were removed from both ships.

In 1942-44 two more of the old 3-in guns, the 20-mm guns and the torpedo tubes were removed; seven 12.7-mm machine-guns and an antisubmarine armament were added. In addition the bridge was rebuilt and the foremast was replaced by a heavy tripod to carry new fire control positions. They ended their service lives as training and depot ships at Callao, where they eventually were broken up after being sold in 1958.

Length: 115.8 m (380 ft) oa *Beam:* 12.3 m (40 ft 6 in) *Draught:* 4.34 m (14 ft 3 in) *Machinery:* 2-shaft 4-cylinder reciprocating, 14000 ihp=24 knots *Protection:* 38 mm ($1\frac{1}{2}$ in) deck over machinery; 76 mm (3 in) conning tower; 76 mm (3 in) shields on 6-in (152-mm) guns *Armament:* 2 6-in (152-mm); 8 3-in (76-mm); 8 6-pdr; 2 1-pdr; 2 18-in (46-cm) torpedo tubes (submerged) *Crew:* 743 *(Grau)*, 746 *(Bolognesi)*

Almirante Latorre

Chilean battleship. Prior to the First World War a minor naval race was in progress among the major naval powers of South America. It had begun in 1907 when Brazil ordered two battleships, *Minas Geraes* and *Sao Paulo*, from Vickers. Argentina countered this move by ordering the *Rivadavia* and *Moreno* from the Fore River yard in 1910. Brazil then ordered another ship, the *Rio de Janeiro*, from Armstrongs, and so Chile decided to construct two vessels of her own and ordered them from the same company. Originally named *Valparaiso* and *Santiago*, they were changed to *Almirante Latorre* and *Almirante Cochrane* while under construction. They were large and powerful ships, and like the majority of South American Dreadnoughts had been ordered with little thought as to the practical problems of manning and running such vessels.

As originally designed they displaced 27400 tons, had a speed of 23 knots and carried an armament of ten 14-in and 22 4.7-in guns. At this time, however, the European powers had been adopting heavier calibre guns for the secondary armament of their Dreadnoughts and so it was decided to replace the 4.7-in guns by 16 6-in. This modification increased the designed displacement to 28000 tons and reduced the speed by $\frac{1}{4}$ knot.

Almirante Latorre was laid down on November 27, 1911 and launched on November 27, 1913. On the outbreak of war in August 1914 all work on her was suspended as British shipyards turned their attention to the more pressing needs of the Royal Navy. On September 9 the ship was purchased by the British Government and work on her recommenced. She was completed in September 1915 and joined the Grand Fleet of the Royal Navy as HMS *Canada*. Her sister ship, *Almirante Cochrane*, not being so far advanced, was not purchased until 1917. She was redesigned as an aircraft carrier, launched in 1918 and completed as HMS *Eagle*.

The *Canada* was sold back to Chile in April 1920 after a refit at Devonport. On August 1 she reverted to her earlier name of *Almirante Latorre*. The ship had of course been slightly modified by the British during her five years with the Royal Navy. Of these modifications

National Maritime Museum

***Almirante Latorre* in 1918 while serving as HMS *Canada*. She was returned to Chile in 1920**

the most important had been the removal of two of the 6-in guns because they were subject to blast from Q turret, the rearrangement of the bridge structure and fire control arrangements and the fitting of additional deck protection over the magazines.

Between 1929 and 1931 she underwent a major refit at Devonport. Her boilers were converted to burn oil fuel only and new Parsons turbines, manufactured by Vickers Armstrong, replaced her original installation. She was also fitted with bulges which increased her beam to 31.4 m (103 ft) and reduced the draught by 30.5 cm (12 in). An aircraft catapult was fitted on the quarterdeck and the two 3-in AA were replaced by four 4-in AA guns, all mounted on the after superstructure.

The *Latorre* served as flagship of the Chilean navy until 1958 when she was placed on the disposal list. In May 1959 she was sold to Japanese shipbreakers and arrived in Japan for scrapping on August 28, 1959.

See also *Canada, Eagle.*

Displacement: 28600 tons (at load draught), 32120 tons (full load) *Length:* 201.5 m (661 ft) (oa) *Beam:* 28 m (92 ft) *Draught:* 8.9 m (29 ft 6 in) (mean) *Machinery:* 4-shaft Brown Curtis and Parsons turbines, 37000 shp=$22\frac{1}{2}$ knots *Protection:* 228 mm (9 in) (max) belt; 254 mm (10 in) (max) gun houses; 254 mm (10 in) (max) barbettes; 114 mm ($4\frac{1}{2}$ in) and 76 mm (3 in) bulkheads; 25-51 mm (1-2 in) decks; 4-in (102-mm) over steering gear *Armament:* 10 (5×2) 14-in (355-mm); 16 6-in (152-mm); 2 3-in (76-mm) AA; 4 21-in (530-mm) torpedo tubes (submerged) *Crew:* 1167

Almirante Oquendo

Spanish antisubmarine destoyer class. The *Oquendos* were originally designed as conventional destroyers, and were intended to carry an armament of 8 105-mm, 6 37-mm and 4 20-mm guns, 7 torpedo tubes (2 double and 1 triple) and two depth-charge throwers. Then they were redesigned and classed as antisubmarine frigates in 1955. They were eventually rated as antisubmarine destroyers in 1961.

Nine of the class were ordered from Ferrol in 1947-48. Of these, *Blas de Lazo, Blasco de Gavery, Bonifaz, Gelmirez, Langare* and *Recalde* were cancelled in 1953. Only three were completed (see chart).

The length of time taken to complete these ships was partly due to Spain's precarious financial position in the 1950s, partly to uncertainties over the role these vessels should play and partly to the need to ensure that they were not obsolete before they even joined the navy.

When completed, they formed two entirely separate sub-groups, the *Almirante Oquendo* having a totally different weapons and electronics fit to the *Roger de Lauria* and the *Marques de la Ensenada*. The *Almirante Oquendo* was first completed in September 1960, and was then immediately taken in hand for modernization to make use of the American aid that had become available as a trade-off for the American use of two airbases and the submarine base at Rota.

When finally completed, she had a British radar outfit, and had two twin 120-mm AA turrets, one forward and one aft.

The *Roger de Lauria* and *Marques de la Ensenada*, which were much less complete, were returned to Cartagena for a complete reconstruction. The beam was increased by two metres to cope with the extra top hamper and the forecastle was lengthened to leave only a short quarter deck. They were fitted with three twin 127-mm (5-in) turrets, two forward and one aft, two triple launchers for antisubmarine torpedoes, and a landing pad and hangar for an antisubmarine helicopter at the break of the forecastle. In addition to an extensive American radar fit, they also have variable depth sonar at the stern, and are the equivalent of the American *Allen M Sumner* FRAM II conversions. Only the machinery is now the same as that in the *Almirante Oquendo.*

(Almirante Oquendo) Displacement: 2050 tons (standard), 2765 tons (full load) *Length:* 116.4 m (381 ft 10 in) *Beam:* 11.0 m (36 ft) *Draught:* 3.85 m (12 ft 7 in) *Machinery:* 2-shaft geared steam turbines, 60000 shp=38 knots *Armament:* 4 120-mm (4.7-in); 6 40-mm (1.5-in); 4 20-mm; 6 torpedo launchers (2 triple); 2 21-in (53-cm) torpedo tubes (fixed above water) *Crew:* 249

Vessel	laid down	launched	completed
Almirante Oquendo	6/1951	9/1956	4/1963
Roger de Lauria	9/1951	11/1958	5/1969
Marques de la Ensenada	9/1951	7/1959	9/1970

Alouette, Aérospatiale

French helicopter. First flown in March 1955, the Sud-Est (later Sud-Aviation, now Aérospatiale) Alouette II (Lark II) was the world's first successful turbine-powered helicopter. It won the French company international military and civil business, and production was continued until 1975 when about 1320 had been delivered to 126 operators in 46 countries. Many Alouette II helicopters are used by military or para-military organizations, and in 1956 they were the first ever to carry and fire antitank wire-guided missiles (SS.10 and SS.11). Engines are the 360-hp Turboméca Artouste IIC, or 530-hp Astazou derated to give 360 hp under all temperature and altitude conditions.

The SA.315B Lama has the engine and rotor system of the bigger Alouette III, and went into production in India as the Hindustan Aircraft Cheetah for the Indian army. The SA.316B Alouette III has a larger fuselage bubble seating pilot and six passengers, a

Alouette II of the Aviation Légère de l'Armée de Terre. The French armed forces have bought a total of 378 Alouettes. *Below:* Alouette II of ALAT armed with Nord AS.11 antitank missiles

ECP Armées

Alouette, Aérospatiale

ECP Armées

monocoque (not lattice) tail boom, and large rotor restressed to match an 870-hp Artouste IIIB derated to a constant 570 hp. Many remain in service carrying various missiles, a 20-mm MG 151/20 cannon or 7.62-mm AAT-52 machine-gun, ambulance gear or other equipment. By 1976 deliveries exceeded 1395.

The SA.319B has the more efficient 870-hp Astazou engine, derated to 600 hp, and has proved especially useful in naval roles carrying air-sea rescue gear, ASW sensors and two Mk 44 torpedoes, or a stabilized sight and radar plus two AS.12 antiship missiles.

(SE.3130 Alouette II) *Rotor diameter:* 10.2 m (33 ft 5½ in) *Fuselage length:* 9.7 m (31 ft 10 in) *Gross weight:* 1600 kg (3527 lb) *Max speed:* 185 km/h (115 mph)

(SA.319B Alouette III) *Diameter of main rotor:* 11.02 m (36 ft 1¾ in) *Length (rotors turning):* 12.84 m (42 ft 1½ in) *Gross weight:* 2250 kg (4960 lb) *Max speed:* 220 km/h (136 mph)

ECP Armées

French Army Alouette III conducting trials with AS.11 wire-guided antitank missiles, four of which can be carried. Other assault versions of the Alouette III mount an internal 7.62-mm AAT-52 machine-gun or 20-mm MG 151/20 cannon or two AS.12 or Hot missiles

Left: Alouette III armed with four AS.11s. Developed from the SS.11 surface-to-surface missile, the AS.11 is a general purpose battlefield missile and can be equipped with various types of warhead, including nuclear depth-charges for use against submarines

Above: The Alouette III, an extrapolation of the Alouette II, with a more powerful 870-hp Astazou turboshaft engine derated to 600 hp and strengthened transmission system, can be used in the tactical or assault transport, flying crane or casualty evacuation roles

Alpha

Russian submarine class. The Russians have always had an interest in underwater warfare. They were the first to use mines on a large scale, and were early exponents of the submarine. Their submarine fleet at the start of the Second World War was the largest —though by no means the most effective—in the world.

After the end of that war they captured many advanced German submarine projects which they made good use of in building up a very large fleet of conventional submarines.

When the US produced the first nuclear submarines, Russia was quick to follow. Her first class of nuclear fleet submarines, the *Novembers*, began construction in 1958 and they first entered service in 1961. Fifteen *Novembers* were built, but they were very noisy. In 1970 one sank near the English Channel.

The next Russian class of fleet submarines, the *Victors*, commenced construction in 1966. They represent a great advance over the *Novembers*.

Both these classes were built in batches, apparently without a prototype.

Then in 1970 a Russian submarine was spotted that was apparently nuclear powered, and by its size and shape it would appear to be a fleet submarine—that is, one that is equipped to hunt other submarines and surface vessels. It was assigned the NATO reporting name of *Alpha*.

By 1976 no other submarine of that class had appeared, and it differs significantly both from the *Victor* Class that preceded it and from the *Uniform* Class of fleet submarines, which first appeared in 1974. It is conceivable that the *Alpha* was a trials ship for a new type of reactor, since all other types of Russian fleet submarine have been built in substantial numbers, without prototypes having appeared before the production boats.

Displacement: 3500/4500 tons (surfaced/submerged) *Machinery:* probably nuclear

Note: Although the Russians assign names or numbers to their submarines, NATO assigns to each class a letter of the standard NATO alphabet when it is first observed, and it is by this reporting name that they are usually known in the West, as in many cases the Soviet names or designations are not known.

Alpha Jet
Dassault-Breguet/Dornier

Franco-German light strike/trainer. The Alpha Jet was the successful contender for a joint Franco-German requirement of 1969 for a new jet trainer. Since that time the design has been greatly developed, and in Luftwaffe service a multi-role aircraft quite different in detail from the French trainer will serve in both the operational training and battlefield close-support and attack roles.

Industrially the main partners are Dassault-Breguet and Dornier for the airframe and SNECMA/Turboméca and Germany's KHD and MTU on the twin 1345-kg (2965-lb) thrust Larzac 04 turbofan engines. The engines are installed on each side under the high down-sloping wing, fed by fuselage cells and integral tanks forming the outer wings. The latter have fixed leading edges but powerful double-slotted trailing-edge flaps; the German partner is modifying one aircraft to have a supercritical wing with flaps on both leading and trailing edges for extremely powerful manoeuvres. The pressurized cockpit accommodates two in tandem seats (Martin-Baker Mk 4 in France, Stencel SIIIS in Germany), the trainer seating the instructor at the rear and the Luftwaffe attack configuration turning the backseater into a *Kampfbeobachter* (combat observer).

No provision is made for an internal gun, tactical sensors or flight refuelling, but much

The Alpha Jet is a joint development by Dassault-Breguet and Dornier, and quantities of quite different versions have been ordered by the Luftwaffe and the Armée de l'Air. The German version (right) is equipped for operational training, reconnaissance and light attack roles, and is armed with a 30-mm DEFA or 27-mm Mauser cannon with 150 rounds of ammunition or two 12.7-mm Browning machine-guns with 250 rounds each in a pod below the fuselage. Four underwing stores stations are each capable of carrying one 113-kg (250-lb), 227-kg (500-lb) or 454-kg (1000-lb) bomb, a 272-kg (600-lb) cluster dispenser or a pod containing 36 70-mm (2.75-in) rockets

The French variant of the Alpha Jet (below and bottom) is equipped solely as an advanced trainer for the Armée de l'Air, to replace Lockheed T-33As and Mystère IVAs

Dassault-Breguet

Dassault-Breguet

can be added externally at the penalty of reduced performance. A fuselage pod can accommodate reconnaissance sensors, a 30-mm DEFA or 27-mm Mauser cannon with 150 rounds, two 12.7-mm Brownings each with 250 rounds, or ECM. Two underwing pylons can carry tanks, and in German aircraft there will be four wing pylons for external mission load up to 2200 kg (4850 lb). The Luftwaffe aircraft have steerable nosewheel, arrester hook, different brakes, different fuel system, a yaw damper, a Kaiser HUD (head-up display) for weapon aiming and, probably, a small Doppler radar.

The first prototype flew on October 26, 1973. Originally it had been expected that France and Germany would each order 200, for delivery in early 1976, but the programme has slipped badly. After protracted multinational discussion a preliminary £160 million contract was signed in January 1976 for 140 aircraft, 56 of them French trainers and the rest Luftwaffe multi-role machines. First delivery will be in the third quarter of 1978. Belgium has ordered 16 trainers, with an option on 17, and discussions have been going on since mid-1975 with the Turkish aircraft company Turk Ucak Sanayii Anonim Sirketi, and with the industry in Egypt, both of which wish to build an aircraft in this class under licence.

Span: 9 m (29 ft 10¾ in) *Length:* 12.25 m (40 ft 3¾ in) *Gross weight:* (trainer) 4890 kg (10780 lb); (close support) 7000 kg (15432 lb) *Max speed:* (clean) 991 km/h (616 mph)

Alpino

Italian frigate class. The *Alpino* Class frigates had their origin in the provision for two frigates to be named *Circe* and *Climene* in the Italian 1959-60 naval programme. Alterations to both armament and machinery were made to the design in 1962, making them the first Italian warships to use gas turbine propulsion. Their names were changed in June 1965, *Circe* becoming *Alpino* and *Climene* becoming *Carabiniere*.

Name	laid down	launched	completed
Alpino	2/1963	6/1967	1/1968
Carabiniere	1/1965	9/1967	1968

Based on the preceding *Bergamini* and *Centauro* Classes, they resemble the *Bergaminis* in appearance apart from a much more prominent funnel, but are larger and more heavily armed. They carry two Agusta-Bell 204 antisubmarine helicopters and are fitted with variable depth sonar.

However, the main difference between the *Alpinos* and their predecessors lies not in their size or armament but in their machinery. In place of the *Centauro*'s steam turbines and the *Bergamini*'s diesels, the *Alpinos* were given a CODAG (Combined Diesel And Gas) turbine arrangement. This consists of four Tosi diesels developing a total of 16800 shp, and two Tosi-Metrovick gas turbines giving 15000 shp.

The diesels can be used on their own for cruising, and give a maximum speed of 22 knots. The gas turbines are used for quick starts and rapid acceleration, and boost the maximum speed by six knots.

Two more of the class were projected, but the *Lupo* Class, a slightly smaller improved version, was ordered instead.

Displacement: 2700 tons (full load) *Length:* 113.3 m (371 ft 7 in) *Beam:* 13.3 m (43 ft 7 in.) *Draught:* 3.9 m (12 ft 10 in) *Machinery:* 2-shaft diesels and gas turbines, 31800 shp=28 knots *Armament:* 6 3-in (76-mm); 1 depth-charge mortar; 6 12-in (30.4-cm) torpedo tubes (2 triple) *Aircraft:* 2 helicopters *Crew:* 254

Alsedo

Spanish destroyer class. The three ships of this class were authorized under a law of 1915 but the first was not laid down until 1920. The *Alsedo* was launched in October 1922, the *Valasco* in June 1923 and the *Lazaga* in March 1924. For recognition purposes the initial letters of their names were painted on the bow. By the time they were completed they were somewhat out of date, being comparable to foreign destroyers constructed in the early years of the First World War. They were small and had a comparatively light armament, but had a good turn of speed and made 36 knots on trial. The 4-in guns were disposed on the centre line, fore, aft and amidships, and the 2-pdr guns abreast the bridge on the forecastle deck. They had four funnels, the foremost being taller than the remainder, and the torpedo tubes were mounted on the centre line between the after funnel and the main mast.

They had comparatively long but uneventful careers. During the Spanish Civil War, on September 19, 1936, the *Valasco* sank the Government submarine *B 6* by gunfire off Ferrol. The *Alsedo* and *Valasco* were removed from the effective list in 1957 and the *Lazaga* in 1961.

Alsedo, Valasco and *Lazaga* built by Soc. Español de Construcción Naval at Cartagena.

Displacement: 1145 tons (normal), 1315 tons (full load) *Length:* 86.25 m (283 ft) *Beam:* 8.23 m (27 ft) *Draught:* 4.57 m (15 ft) *Machinery:* 2-shaft geared steam turbines, 33000 shp=34 knots *Armament:* 3 4-in (102-mm) (3×1); 2 2-pdr (2×1); 4 21-in (53-cm) torpedo tubes (2×2) *Crew:* 70

Alvis British tanks — See **Scorpion**

AM-1 US aircraft — See **Mauler**

AM.39 French ssm — See **Exocet**

Amatsukaze

Japanese destroyer class. At the outbreak of war in August 1914 Japan was very short of fleet escort destroyers, so a new construction programme was hurriedly inaugurated. Under this the *Amatsukaze* Class were laid down as part of the Additional Reinforcement Programme of 1915. They were rated as first class ocean-going destroyers.

To save time the design was based on the *Yamakaze*—the first Japanese ocean-going destroyer, built 1910-11—though opportunity was taken to incorporate modifications based on experience. Information obtained from the *Urakaze*—built by the British firm Yarrow and completed in September 1915—was also worked into the design.

There were four ships in the class:

Name	ordered	laid down	completed
Amatsukaze	1916	10/1916	4/1917
Hamakaze	1916	10/1916	2/1917
Isokaze	1916	10/1916	3/1917
Tokitsukaze	1916	12/1916	5/1917

As is the case with many Japanese destroyers built up to 1945, they were named after winds. *Amatsukaze* itself means 'Heavenly Wind'.

At this time the Japanese were concerned about the ever-growing size of their destroyers, and an attempt was made to keep the displacement of the *Amatsukazes* down by using longitudinal construction. Destroyers then were normally built with transverse frames on the keel connected by longitudinal girders. In the *Amatsukazes*, long frames parallel to the keel were connected by transverse members. This method of construction gives a lighter hull, but although it was used for larger ships it was not a success in the *Amatsukazes*, and subsequent Japanese destroyers returned to the transverse method of construction.

The *Hamakaze* introduced another innovation to the Japanese navy. In 1916 the Imperial Japanese Navy completed a ship model testing tank and the Mitsubishi Company, who were about to lay down the *Hamakaze*, took advantage of this to improve her hull form. A model of *Hamakaze* was tested—the first time a ship of the Imperial Japanese Navy built in Japan was tested in this way.

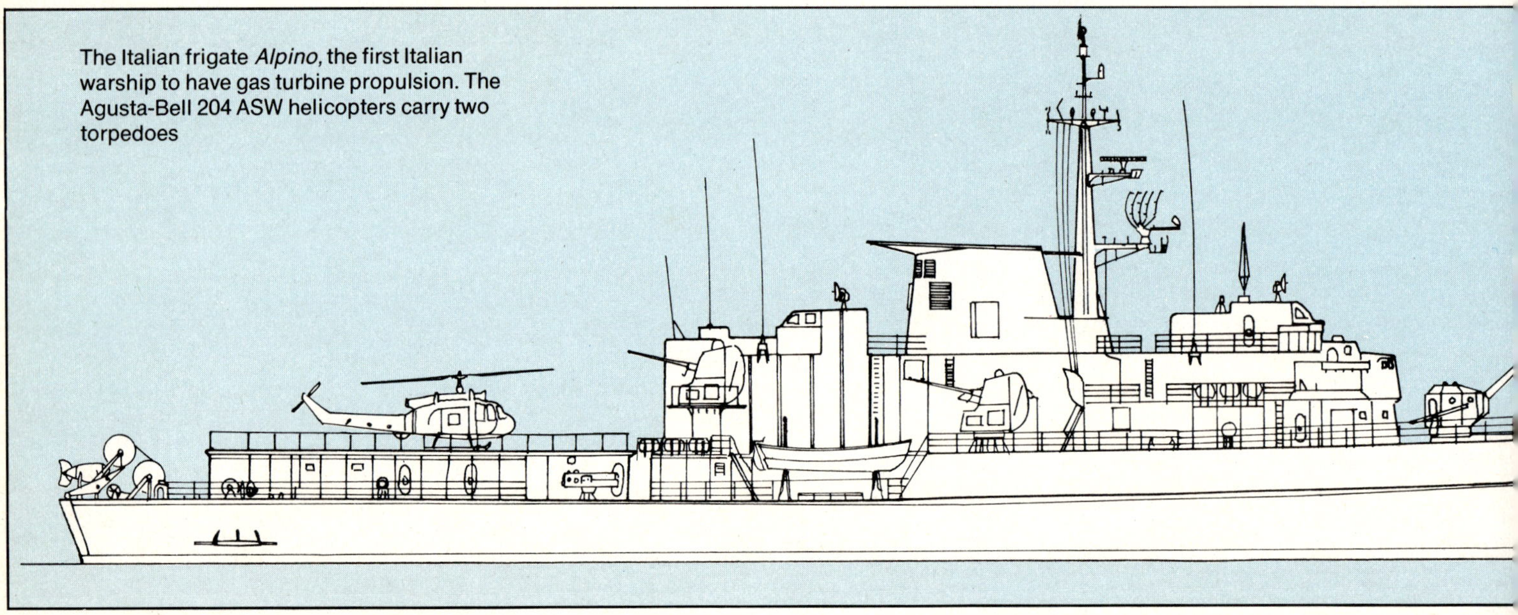

The Italian frigate *Alpino*, the first Italian warship to have gas turbine propulsion. The Agusta-Bell 204 ASW helicopters carry two torpedoes

The British Type 21 frigate HMS *Amazon*, the first of a class of eight built since 1969, based on a Vosper Thornycroft commercial design

Unlike *Urakaze*, the *Amatsukazes* were not fitted with 21-in torpedo tubes. Not until their successors *Kawakaze* and *Tanikaze* did 21-in torpedo tubes become standard on Japanese destroyers.

The *Amatsukazes* had relatively uneventful careers, and were all discarded in 1935. *Amatsukaze*, *Hamakaze* and *Isokaze* were scrapped almost immediately. *Tokitsukaze* was used as a hulk, and was not finally broken up until 1948.

Displacement: 1105 tons (normal), 1227 tons (full load) *Length:* 96.68 m (317 ft 1 in) *Beam:* 8.52 m (27 ft 11 in) *Draught:* 2.83 m (9 ft 3 in) *Machinery:* 3-shaft steam turbines, 27000 shp=34 knots *Armament:* 4 120-mm (4.7-in); 2 7.7-mm; 6 18-in (46-cm) torpedo tubes (3 twin above water) *Crew:* 128

Amatsukaze

Japanese destroyer. In 1945 the Imperial Japanese Navy ceased to exist, and it was not until the Korean war that the Maritime Self Defence Force (MSDF), as the Japanese navy is now known, came into being.

The terms of the Japanese constitution preclude the use of ships designed for offensive purposes, and for many years this meant that the largest ships in the MSDF were destroyers and frigates of less than 3000 tons full load.

However, in 1960 a large antisubmarine escort named *Amatsukaze* (Heavenly Wind) was ordered. Laid down in November 1962 and launched in October 1963, she was completed in February 1965. Fitted with American weapons and electronic equipment (as are all MSDF warships) she was the first Japanese vessel to be fitted with guided missiles (a single Tartar medium range antiaircraft missile launcher aft) and also the first to carry and operate a helicopter for antisubmarine warfare.

When completed, she was the largest Japanese warship built since the war, displacing 1100 tons more than the next largest, the destroyers *Akizuki* and *Teruzuki*, though the considerably larger helicopter-cruiser *Haruna* (which has hangar space for several helicopters) was later taken into service.

Unlike the Americans, whose warships need to carry a large amount of fuel for operations overseas, MSDF vessels are only required to operate in home waters. The weight and space saved by the smaller amount of fuel carried was used to provide a heavier armament, *Amatsukaze* being no exception. As well as the single Tartar launcher aft she mounts two twin 3-in turrets and two Hedgehogs forward, and antisubmarine dropping gear amidships. In addition she can operate an antisubmarine helicopter.

However, no more ships of this type were constructed, the Japanese considering that *Haruna*, a considerably bigger ship with a large helicopter platform and on-board maintenance facilities, represents a better investment.

The present *Amatsukaze* is not the first ship of that name to introduce new concepts to the Japanese navy. The *Amatsukaze* of 1917 was the first Japanese destroyer to have longitudinal framing, and the *Kagero* Class destroyer *Amatsukaze* of 1939 was the first to have high pressure boilers.

Displacement: 3050 tons (standard), 4000 tons (full load) *Length:* 131 m (429 ft 8 in) *Beam:* 13.4 m (43 ft 11 in) *Draught:* 4.2 m (13 ft 9 in) *Machinery:* 2-shaft geared steam turbines, 60000 shp=33 knots *Armament:* 1 single Tartar missile launcher; 4 3-in (76-mm); 2 Hedgehog; 2 torpedo dropping racks (one on each beam above water) *Aircraft:* 1 helicopter *Crew:* 290

Amazon

British frigate class. In the 1960s the Royal Navy planned to follow its very successful Improved Type 12 *Leander* Class general-purpose frigates with a new design carrying an antisubmarine helicopter and the Seawolf short-range antiaircraft missile. This new design evolved into the Type 22 *Broadsword* Class. However, by the late 1960s it had become obvious that the extended development of Seawolf and several other systems intended for the Type 22s meant that they could not be in service before the mid-1970s at the earliest.

The first *Leander* was laid down in 1959, and the basic design dated back to 1951, when the first *Whitby* was ordered. The hull could no longer accommodate all the weapons systems and electronic equipment that were desirable. Therefore the last of the 26 *Leanders* was completed in 1972, and a stopgap design had to be found to fill the gap before the Type 22s came into service.

Fortunately, a suitable design was at hand. The commercial shipbuilders Vosper Thornycroft had proposed a series of designs for vessels ranging in size from corvettes to large guided-missile destroyers, all of which shared the same basic layout, hull-form and appearance. Four variants of the Mark V design, the 1100-ton *Saams*, were already under construction for Iran.

Vosper Thornycroft was therefore given a contract on February 27, 1968, to design a patrol frigate in collaboration with Yarrow, based on the Vosper Thornycroft commercial frigate designs. This was the first time the

British Admiralty had adopted a commercial design since the Thornycroft 'Hunt' Type IV escort destroyers *Brecon* and *Brissendon* of 1942. That the procedure was justified is shown by the fact that the first of the *Amazon* Class was laid down only 20 months after the design contract was placed.

There are eight in the class. The first three were built by Vosper Thornycroft, and the last five by Yarrow.

They are handsome ships, with a 4.5-in Vickers Mk 8 gun forward, a single 20-mm gun either side of the bridge for counter-insurgency work, a quadruple Seacat antiaircraft missile launcher over the hangar at the break of the quarterdeck, and a landing pad for the Lynx ASW helicopter right aft.

Arrow was the first of the class to be fitted with two MM38 Exocet surface-to-surface missiles, mounted on the break of the forecastle, just under the bridge. The first four vessels were to be refitted with Exocet at a later date, and the entire class was also to have two triple 21-in torpedo launchers as they became available.

The *Amazons* were the first frigates to be designed and constructed with COGOG machinery (COmbined Gas Or Gas), an arrangement tried out in the Converted *Blackwood* frigate *Exmouth,* and also fitted in the Type 42 *Sheffield* Class missile destroyers. The machinery consists of two Tyne gas turbines for cruising and two Olympus for maximum speed.

Displacement: 2500 ton (full load) *Length:* 117 m (383 ft 10 in) *Beam:* 12.7 m (41 ft 8 in) *Draught:* 3.7 m (12 ft 2 in) *Machinery:* 2-shaft gas turbines (COGAG) 50 000 shp =30+ knots *Armament:* 1 4.5-in (115-mm); 2 20-mm; 2 Exocet; 1 Seacat quadruple launcher *Aircraft:* 1 Lynx antisubmarine helicopter *Crew:* 170

Name and Pendant No	laid down	launched	completed
Amazon F169	11/1969	4/1971	5/1974
Antelope F170	3/1971	3/1972	2/1975
Active F171	7/1971	11/1972	1976
Ambuscade F172	7/1971	1/1973	9/1975
Arrow F173	6/1972	2/1974	6/1975
Alacrity F174	2/1973	9/1974	
Ardent F175	2/1974	5/1975	
Avenger F176	10/1974	11/1975	

The Royal Navy's Type 21 frigates *Amazon* (F169) and *Antelope* (F120) meet in the English Channel, January 1975, during the *Antelope*'s sea trials. The *Amazon* Class were built as a stopgap during the extended development of the *Broadsword* Class. Armed with an automatic quick-firing 4.5-in Vickers Mk 8 gun and a quadruple Seacat close-range antiaircraft and surface-to-surface missile launcher, and with accommodation for a Lynx ASW helicopter, the first four ships are to have Exocet launchers added. Two Tyne and two Olympus gas turbines give the ships a maximum speed of over 30 knots

Amazon/Ambuscade

British destroyers. Ordered under the 1924-25 Estimates, the *Amazon* and *Ambuscade* were the first British destroyers to be laid down after the First World War. Constructed as experimental vessels, their design was based on war experience and incorporated the latest developments in machinery. They were the first British destroyers to employ boilers which produced superheated steam. On trials the *Amazon* made 37.5 knots with 41 460 shp and *Ambuscade* 37.16 knots with 32 770 shp. The *Amazon* was designed and built by Thornycroft and the *Ambuscade* by Yarrow; both were launched in 1926 and completed in 1927. The design of the 'A' to 'I' Class destroyers was based on the experience gained with these two vessels.

They spent the majority of their active wartime careers on escort duties and were specially modified for antisubmarine duties in 1942-43. Early alterations included the substitution of a 3-in AA gun for the after bank of torpedo tubes. In 1942 *Amazon* had 'A' gun replaced by a Hedgehog antisubmarine weapon and 'Y' gun removed to make more space for depth charges. *Ambuscade* was similarly modified in 1943 but was fitted with a Squid instead of the Hedgehog. *Amazon* was taken in hand again in 1943 and had the forward bank of torpedo tubes and the 3-in AA gun removed to provide for a further increase in the depth-charge armament.

On April 29, 1940, *Amazon* assisted in the sinking of the submarine *U 50* in the North Atlantic. In 1944 she became an aircraft target and in January 1945 was placed in Reserve. In 1948 she was sold to Arnott Young and arrived at Troon in April 1949, where she was broken up. In January 1943 the *Ambuscade* became an air target for the

MOD

Wright & Logan

Above: **HMS *Amazon,* one of the pair of experimental British destroyers built in the 1920s which formed the basis for the subsequent 'A' to 'I' Classes. *Below:* USS *America* off Cannes, France, with Phantoms, Skyhawks, a Vigilante, Trackers and a Sea Sprite helicopter on her deck**

US Navy

Fleet Air Arm and during 1944 and 1945 was employed as an antisubmarine training ship. She was sold to Arnott Young in November 1946 and arrived for breaking up in 1947.

(Amazon) Overall length: 98.4 m (323 ft) *Beam:* 9.6 m (31 ft 6 in) *Draught:* 2.81 m (9 ft 3 in) *Standard displacement:* 1352 tons *Armament:* 4 4.7-in (119-mm); 2 2-pdr; 2 triple 21-in (53-cm) torpedo tubes *Machinery:* Brown Curtis geared turbines with Parsons geared turbines for cruising. Three Yarrow boilers driving two shafts. *Shaft horsepower:* 42 000 *Speed:* 37 knots *(Ambuscade) Overall length:* 98.1 m (322 ft) *Beam:* 9.45 m (31 ft) *Draught:* 2.6 m (8 ft 6 in) *Standard displacement:* 1173 tons *Armament:* as *Amazon Machinery:* as *Amazon Shaft horsepower:* 35 500 *Speed:* as *Amazon*

America

US aircraft carrier. After the first two *Kitty Hawk* Class aircraft carriers had been laid down it was decided that all future American carriers would be powered by nuclear reactors. Next, USS *Enterprise* was given a nuclear powerplant, but she was so expensive to build that Secretary of Defense McNamara abandoned nuclear power for the next two, *America* (CVA66) and *John F Kennedy* (CVA67). These reverted to oil-fired boilers and were built to virtually the same design as the *Kitty Hawk* and *Constellation.*

America was built at Newport News, and was laid down in January 1961. Launched in February 1964, she was completed in January 1965 at a cost of $248 800 000, a fast building time for such a large and complicated ship.

She is fitted to carry between 70 and 90 aircraft, depending on the size of those embarked. Until recently she was rated as an Attack Aircraft Carrier (CVA) and carried only fighter and attack squadrons, their associated reconnaissance and early warning aircraft and rescue helicopters. Now that the wartime *Essex* Class carriers, which were modified for antisubmarine work, have been scrapped, the CVAs have had to embark antisubmarine aircraft. This has meant a diminution in the number of fighter and attack aircraft carried, and the carriers have been rerated as CVs.

The carrier herself is armed with two Mark 10 Twin Terrier medium range antiaircraft missile launchers—an updated version which can also accommodate Standard. She has a very extensive radar fit and the Naval Tactical Data System. *America* was the first American carrier to be fitted with sonar.

She saw active service off North Vietnam, and has also been in the US 6th Fleet. During the 1967 Arab-Israeli war, she launched her planes when Israeli aircraft and torpedo boats attacked the USS *Liberty.*

Displacement: 60 300 tons (standard) 78 250 tons (full load) *Length:* 319.3 m (1046 ft) oa *Beam:* 39.6 m (129 ft 11 in) (hull), 76.8 m (252 ft) (flight deck) *Draught:* 10.9 m (35 ft 9 in) *Machinery:* 4-shaft geared steam turbines, 280 000 shp=35 knots *Armament:* 2 twin Terrier launchers *Aircraft:* 70-90 *Crew:* 2700+2000 aircrew

Amiens Airco D.H.10

British bomber. Better known as the D.H.10, the Amiens was an Airco product, designed by Capt Geoffrey de Havilland, and a more powerful development of the D.H.3. The latter was a twin-engined pusher of 1916, and the same layout was followed in the first Amiens of March 1918, with 240-hp BHP engines. The Amiens II had tractor 360-hp Rolls-Royce Eagle VIII engines, but for the production machine the engines were changed to the 400-hp Liberty 12, and the two wheels under the nose were removed.

The resulting Amiens III was faster than the D.H.9A and despite having a crew of three and two pairs of Lewis guns carried more than twice the bomb load—454 kg (900 lb)—mainly inside the square-section fuselage. Large orders were placed, but the Armistice came in November 1918 just before the Independent Air Force Amiens squadrons, led by No 104 Squadron, began operations. Contracts were cancelled at the 220th aircraft. The last batches had the engines moved down to rest on the lower wing, and were designated Mk IIIA.

In 1919 several Amiens flew courier and mail services in Europe, especially between Hawkinge and Cologne. On May 14-15, 1919, one aircraft on this route became the first to carry mail by night anywhere in the world. The main operator in Europe was RAF 120 Squadron. In Egypt 216 (Bomber Transport) Squadron pioneered the Cairo-Baghdad route, and first used the furrow ploughed across the featureless desert as a guide. In India No 60 Squadron's Amiens were in hectic action in 1920 and again in 1922 against rebel tribesmen on the wild north-west frontier of India. The Amiens was withdrawn from RAF service and replaced, usually by the Vimy, in 1924.

(Amiens IIIA) *Span:* 19.96 m (65 ft 6 in) *Length:* 12.07 m (39 ft 7 in) *Gross weight:* 4082 kg (9000 lb) *Maximum speed:* 202 km/h (126 mph)

Musée de l'Air

First production version of the Amiens, the Mk IIIA (D.H.104A), with the engines mounted directly on top of the lower wing

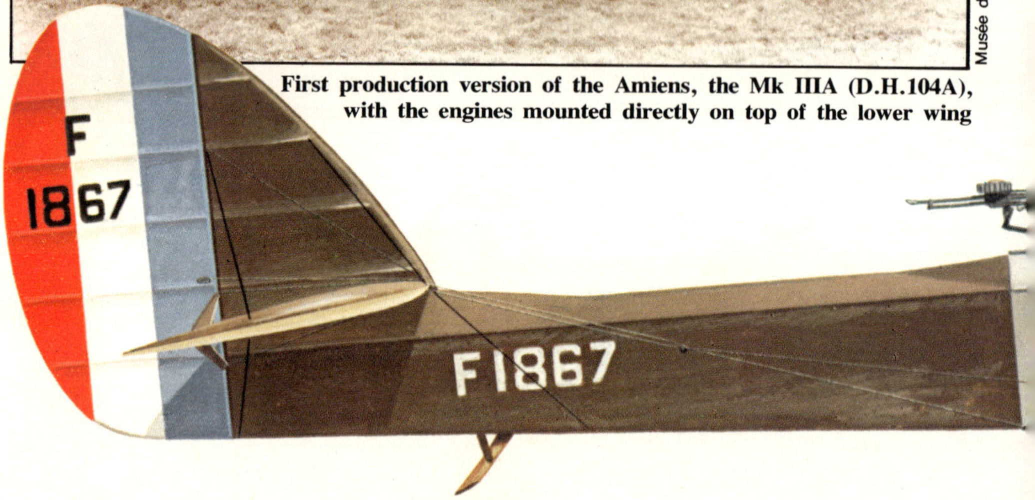

Amiot 143

French bomber. Unlike most of the ugly and indifferent French multi-engine combat aircraft built between the World Wars the 143 saw a lot of action in the spring of 1940. Its origin was a 1928 Multiplace de Combat specification, for day and night bombing, reconnaissance and escort. The Amiot 140, flown in April 1931, beat three other prototypes and led, via the Hispano-powered 142, to the 143, powered by two 740-hp Gnome-Rhône 14Krsd two-row radial engines.

In November 1933 an initial order for 30 was placed. These were designated BCR (Bombardement, Chasse, Reconnaissance) but were destined to serve as five-seat night bomber and reconnaissance machines. Features included a large and roomy fuselage littered with windows, turrets and gun positions (originally fitted with Lewis but eventually carrying a total of four 7.5-mm MAC 1934 machine-guns), an extremely thick wing with no flaps but long narrow ailerons, a mainly unstressed skin of light alloy throughout, fixed spatted landing gear, and a bomb load of up to 900 kg (1984 lb) in racks on the left side of the centre fuselage. They were solid, lumbering, reliable and quite popular aircraft, which on two 900-hp GR14

A flight of Amiot 143s, one of the Armée de l'Air's main bomber types in 1939

Musée de l'Air

The Mk 111, first production version of the Amiens, entered service just too late to see action during the First World War. Although faster, more heavily armed and carrying a heavier bombload than the D.H.9A, the Armistice limited Amiens production to 223

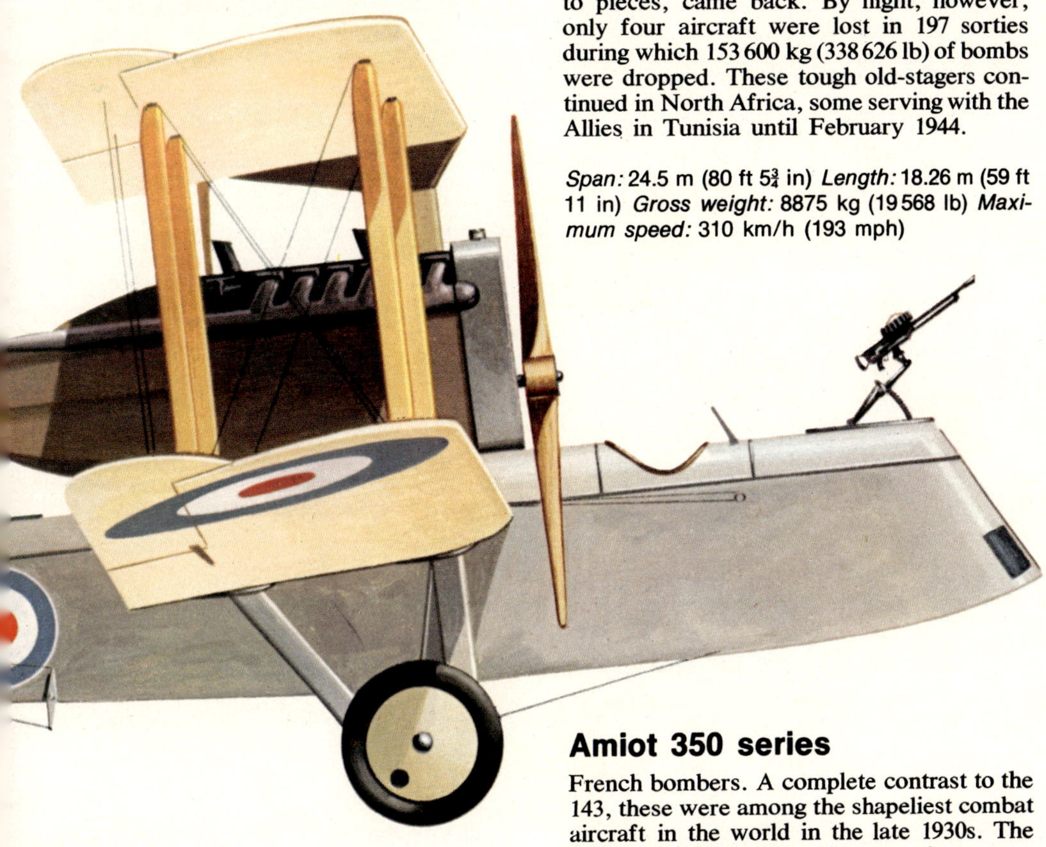

Kirs/Kjrs Mistral Major engines had a performance that still looked good in 1935 (when deliveries began to GB III/22 at Chartres) but proved suicidal in the face of the Bf 109 five years later.

By 1937 a total of 138 Amiot 143s had been delivered, but several later prototypes, mostly with more powerful engines and retractable main gears, had failed to secure orders. The Amiot could defend itself against attack from any direction (unlike the early Wellington, with its power turrets unable to fire to the beam), but was a sitting duck for modern fighters flown with determination. By 1940 about 60 were still in use, with GB I/34, II/34, II/35, I/38 and II/38. After arduous ten-hour leaflet raids through the winter they plunged into heroic bombing missions on May 10, 1940.

The classic mission was the daylight attack by GB I/38 and II/38 on the Sedan bridges from 750 m (2500 ft). A single survivor, shot to pieces, came back. By night, however, only four aircraft were lost in 197 sorties during which 153 600 kg (338 626 lb) of bombs were dropped. These tough old-stagers continued in North Africa, some serving with the Allies in Tunisia until February 1944.

Span: 24.5 m (80 ft 5⅝ in) *Length:* 18.26 m (59 ft 11 in) *Gross weight:* 8875 kg (19 568 lb) *Maximum speed:* 310 km/h (193 mph)

The Amiot 143 was a sturdy and reliable night bomber, but its lumbering performance made daylight missions suicidal

Amiot 350 series

French bombers. A complete contrast to the 143, these were among the shapeliest combat aircraft in the world in the late 1930s. The original SECM-Amiot 341 was a long-range mailplane, exhibited at the Paris Salon of 1936, but it eventually flew in December 1937 as the 340-01 three-seat bomber. The following year, painted to look like a regular service type, it averaged 440 km/h (273 mph) from Villacoublay to Berlin taking Gen Vuillemain, Chief of Staff of the Armée de l'Air, on an official visit to the Luftwaffe.

Eventually many prototypes were built, some with Merlin engines, but ultimately the main production types were the four-seat bombers Amiot 351 and 354, with close-cowled GR 14N two-row radials of 950 or 1060 hp. Beautifully streamlined, these aircraft had a power-aimed Hispano 404 cannon at the rear of the long dorsal canopy and manually aimed MAC 1934 machine-guns in the nose and ventral positions. Bomb load was up to 1200 kg (2645 lb), carried in the fuselage.

Deliveries ran extremely late, but GB I/21, II/21, I/34 and II/34 had received about 85 aircraft by the time the German army reached the Le Bourget plant (aircraft No 87 flew out with minutes to spare and reached Oran). The few sorties that were flown showed the 351 and 354 to be fine aircraft, and only three were lost in action, though a further ten were lost in training and through Luftwaffe attacks on airfields. Subsequently these and related high-speed Amiots were much used by the Vichy forces and Air France during the German occupation.

(Amiot 354 B4) *Span:* 22.83 m (74 ft 11 in) *Length:* 14.5 m (47 ft 6¾ in) *Gross weight:* 11 300 kg (24 912 lb) *Maximum speed:* 484 km/h (301 mph)

Amiral Aube

French armoured cruiser class. The *Amiral Aubes* belonged to a group of 19 very similar French armoured cruisers that commenced with *Jeanne d'Arc* (laid down 1896) and ended with *Waldeck Rousseau* (laid down 1906). All of them were built to much the same design, with four to six funnels, prominent mechanical ventilators and very high silhouettes.

They were built at the instigation of Admiral Fournier, who was imbued with the ideas of the 'Jeune Ecole' (*Amiral Aube* was named after its founder). Fournier intended them to be up-dated and enlarged versions of *Dupuy-de-Lôme.* They were meant for commerce-raiding, and were built instead of, rather than as a supplement to, battleships.

There were five members of the *Amiral Aube* Class:

Name	laid down	launched	completed
Amiral Aube	8/1900	5/1902	1904
Condé	3/1901	3/1902	1904
Gloire	9/1899	6/1900	1904
Marseillaise	1/1900	7/1900	1904
Sully	1900	1901	1905

They cost an average of £875 000 per ship, and were a poor investment for the French navy. Problems with the mechanical ventilation system made them difficult to steam, reducing their already inadequate maximum speed, and, as with all French warships of this period, their guns had a poor performance. British and German weapons of equivalent size could pierce thicker armour and had better accuracy at long range. In addition, they took twice as long to build as their British counterparts.

Sully was lost almost as soon as she had commissioned. She ran aground in 1905 and became a total loss. In the First World War most of the rest of the class served together for much of the time.

Amiral Aube, Gloire and *Marseillaise* (flagship) formed the First Armoured Cruiser Division of the Second Light Squadron, which patrolled the Western Approaches and the English Channel in 1914 and early 1915, and from mid-1918 *Condé*, *Gloire* and *Marseillaise* escorted US troop convoys across the Atlantic. *Condé* spent the first years of the war in the West Indies; *Marseillaise* was stationed there in 1916-17. *Gloire* spent the middle years of the war in the South Atlantic; *Amiral Aube* was in the Mediterranean.

In 1919, *Gloire* escorted General Pershing back to America, and she and *Amiral Aube* were discarded in 1922. *Marseillaise* was used as a gunnery school until she was discarded in 1929, but *Condé*, though discarded in 1933, survived until the end of the Second World War, first as an accommodation ship, then, after capture by the Germans in 1940, as a submarine depot ship at the naval base at Lorient.

Displacement: 9856 tons (normal), 10400 tons (full load) *Length:* 140 m (459 ft 2 in) *Beam:* 20.2 m (66 ft 3 in) *Draught:* 7.70 m (25 ft 2 in) *Machinery:* 3-shaft steam triple-expansion, 20500 ihp=21 knots *Protection:* 170 mm (6.7 in) belt; 45 mm (1.8 in) deck; 200 mm (7.8 in) main turrets *Armament:* 2 194-mm (7.6-in); 8 164.7-mm (6.5-in); 6 100-mm (3.9-in); 18 47-mm (1.85-in); 5 45-cm (17.7-in) torpedo tubes (2 beam submerged; 2 beam, 1 stern, above water) *Crew:* 612

Amiral Bourgois

French submarine. The French navy was the first to employ submarines on a large scale, and in the late 1890s and early 1900s was in the forefront of their development. Yet this early promise was not sustained.

Although French designers continued to be as inventive as before, the French shipbuilding and engineering industry proved incapable of translating their ideas into viable warships. French submarines were continually plagued by poor electrical equipment and periscopes, and the designed power could not be extracted from their diesel engines.

Part of the problem was the lack of adequate scientifically controlled tests. *Amiral Bourgois* (and all other French submarines laid down before 1911) had three pairs of hydroplanes, whereas the US and British had discovered that two were sufficient.

The French were well aware of the shortcomings of their submarines and made determined attempts to overcome them. Provision was made in their 1906 shipbuilding programme for four experimental submarines —*Archimède, Charles Brun, Mariotte* and *Amiral Bourgois*—each of which had a different designer.

Archimède was an improved version of the *Pluviose* Class steam-powered submarines which had been laid down in 1906, and was designed by M Hutter. *Charles Brun,* designed by M Mauriac, was a smaller steam-powered submarine. *Mariotte,* designed by M Radiguer and fitted with Sautter Harle diesels, was the last French 'Sousmarin' and had little surface buoyancy and poor sea-keeping qualities. *Amiral Bourgois* was fitted with Schneider diesels, and had a relatively long range. Her designer was M Bourdelle.

She was built at Rochefort dockyard. Laid down in May 1908, launched in November 1912, and completed in August 1914. This long building time was typical of French warships, and did little to help the French keep abreast of the latest technical developments.

The two best points in the *Amiral Bourgois* design were the long range and the lack of external torpedo racks. Whereas other French submarines of this period had a surface range of 2400-3200 km (1500-2000 miles), *Amiral Bourgois* was capable of 4000 km (2500 miles) at 10 knots.

The absence of Drzewiecki external launching racks meant that her diving depth was not restricted to the torpedo's limit of 30 m or so, as were most other French submarines.

Amiral Bourgois spent the war patrolling in the English Channel, and was discarded in November 1919 after only five years' service.

Displacement: 550 tons (surface); 735 tons (submerged) *Length:* 56.2 m (184 ft 4 in) *Beam:* 5.5 m (18 ft) *Depth:* 3.63 m (11 ft. 10 in) *Machinery:* 2-shaft diesels, 1400 bhp=13.85 knots (surfaced); 2-shaft electric, 1000 shp=8.65 knots (submerged) *Armament:* 4 45-cm (17.7-in) torpedo tubes (internal, 2 forward, 2 aft) *Crew:* 25

Amiral Charner

French cruiser class. The 'Jeune Ecole' of French naval officers, inspired by Admiral Aube, greatly influenced French warship design in the last quarter of the nineteenth century.

The first cruiser to reflect their ideas was *Dupuy-de-Lôme*, and she was soon followed by the four *Amiral Charners*, built to a similar but smaller design at a cost of circa £350000 each.

Name	laid down	launched	completed
Amiral Charner	6/1889	3/1893	1895
Bruix	1890	1894	1896
Chanzy	1890	1/1894	1896
Latouche-Treville	1889	1894	1896

They had the distinctive appearances that all French armoured cruisers of this period possessed, with an exaggerated plough-bow and pronounced tumble-home on the sides. The plough-bow was intended to increase the buoyancy of the forepart without increasing the weight, but in anything of a sea it dug into the waves and made the ships very wet.

This was not a good design. It was too small and the speed was too low for commerce raiding. The complete armour-belt meant that the ships were only very thinly armoured all over. All except *Bruix* had difficulty steaming. In addition, the length of time they took to build meant that they were outclassed by more modern British cruisers. Their best feature was their low silhouette.

Chanzy was lost in 1907, and by 1914 the rest were in reserve or used as training ships. All three returned to active service during the First World War.

Amiral Charner and *Bruix* were first used to escort troop transports from Morocco. In September 1915 *Amiral Charner* rescued 3000 Armenians from the Turks at Antioch Bay. She was torpedoed by *U 21* near Beruit on February 8, 1916, and went down fast. Only one of the 500 men on board survived.

Bruix was employed in the Red Sea in 1915, and in late 1916 and early 1917 took part in the operations against Greece. She ended the war in reserve at Salonika, and was scrapped soon after.

Latouch-Treville was also used in the western Mediterranean in the first months of the war. Then in 1915 she was part of the French fleet at Gallipoli, where she was damaged by Turkish shore batteries. After repairs, she joined *Bruix* off Greece, and ended the war as a training ship for gunlayers. Condemned in 1920, her hull was used in 1925 to raise the battleship *Liberté*, which had blown up in Toulon in September 1911.

Displacement: 4700 tons (normal) *Length:* 110 m (360 ft 10 in) *Beam:* 14 m (45 ft 11 in) *Draught:* 6 m (19 ft 8 in) *Machinery:* 2-shaft steam triple expansion, 8300 ihp=18.2 knots *Protection:* 90 mm (3.5 in) belt; 65 mm (2.5 in) deck; 100 mm (3.9 in) turrets *Armament:* 2 194-mm (7.6-in); 2 138.6-mm (5.4-in); 4 65-mm (2.5-in); 8 47-mm (1.8-in) *Crew:* 375

Ammiraglio Cagni

Italian submarine class. The four large ocean-going submarines of this class were constructed under the 1938 and 1939 Estimates and completed during 1941. They differed substantially from contemporary Italian submarines in that they were specifically designed for long-range operations against merchant vessels. The standard Italian torpedo armament at this time was eight 21-in tubes with one reload for each tube. In the *Cagni* Class the torpedo calibre was reduced to 18-in and the torpedo stowage increased to 36 in order to ensure thay they would not expend their torpedoes too quickly during extended operations. This provided three reloads for each of the forward tubes and two reloads for each of the after tubes. The smaller torpedo, although of less destructive power, was considered of sufficient size for use against merchant ships. They had a substantial range, of 21700 km (13500 miles) at nine knots, and *Cagni* operated in the Atlantic as well as the Mediterranean.

When first completed, being large vessels with a substantial cargo-carrying capacity, they were employed to carry much-needed supplies from Taranto to the North African port of Bardia. *Saint Bon* and *Cagni* carried out their first such operation in October 1941 and in the following month were joined by *Millo* for the second operation. The last members of the class, *Caracciolo,* joined the supply run in December 1941 but had an extremely short career. She was sunk by the British escort destroyer *Farndale* on the return journey from Bardia on December 11. On the 5th of the following month the *Saint Bon* was torpedoed and sunk off Sicily by the British submarine *Upholder.* The *Millo* suffered the same fate in the same area on March 14, 1942, being sunk by one of *Upholder*'s sister ships, the *Ultimatum.*

The *Cagni* had better luck; having left the supply run she operated in the South Atlantic between 1942 and 1943, sinking two small merchantmen. She entered Bordeaux on February 20, 1943, having spent four and a half months at sea. She sailed again in June and operated in the Atlantic until the Italian surrender in September, sinking one merchantman and damaging the British AMC *Asturias.* She gave herself up at Durban on September 20, and was under British operational control for the remainder of the war. She returned to Taranto in January 1944 and served as a training vessel until the end of the war. In February 1948 she was decommissioned and was scrapped at Taranto.

Displacement: 1680 tons (surfaced) 2170 tons (submerged) *Length:* 88.80 m (288 ft 4 in) *Beam:* 7.77 m (25 ft 6 in) *Draught:* 5.18 m (17 ft) *Machinery:* 2-shaft diesel, 4600 bhp=17 knots (surfaced); electric motors=9.1 knots (submerged) *Armament:* 2 3.9-in (100-mm) (2×1); 4 13.2-mm mg (2×2); 14 18-in (46-cm) torpedo tubes (six forward, two aft) *Crew:* 85

Ammiraglio di Saint Bon

Italian battleship class. Italian battleships of the 1870s and 1880s were all built to very idiosyncratic designs. The common features of all the designs were their speed, size, and heavy armament, and their relatively poor armour protection.

After the *Re Umberto* of 1884, Italy did not lay down another battleship until 1893, when the two *Ammiraglio di Saint Bon* Class were commenced.

Name	laid down	launched	completed
Ammiraglio di Saint Bon	7/1893	4/1897	9/1900
Emanuele Filiberto	10/1893	9/1897	4/1900

These were—for Italy—a surprisingly conventional design of light battleship, based on the contemporary British layout of armament and armour. They were armed with two pairs of 10-in guns mounted at either end of an armoured citadel, with a 4-in belt running from the citadel to both bow and stern. This gave a complete waterline armour belt. The 6-in secondary armament was mounted in casemates in the central citadel.

British influence was to be found in all Italian pre-First World War battleships, which also made widespread use of British components. These were either bought from Britain or made in Italy to British designs. The *Saint Bons'* machinery was built in Italy to the designs of Maudsley, Sons and Field, a Thames-side marine engineering firm that engined a number of British battleships.

The *Saint Bons* had a very distinctive appearance, with a funnel set very near either end of the citadel, and a single mast placed halfway between. They cost only £700000 each, but the combination of light armament, heavy armour and moderate speed was unsuitable for Italian requirements, and the design played little part in Italian battleship evolution. However, it did form the basis for a very successful type of Italian armoured cruiser, the first of which, *Giuseppe Garibaldi* (later the Argentine *General Garibaldi*), was laid down in 1894.

The *Saint Bons* first saw action in the Turkish-Italian war of 1911-1912. With the *Benedetto Brin* and *Regina Margherita* they formed the Third Division of the Second Squadron. *Emanuele Filiberto* took part in the bombardment of Tripoli in October 1911, and in April 1912 bombarded the Dardanelles along with the *Saint Bon.*

During the First World War they were both stationed in the northern Adriatic, where they could defend Venice and bombard the Austrian army. *Emanuele Filiberto* was discarded in March and *Saint Bon* in June 1920.

Displacement: 9800 tons (normal) *Length:* 111.8 m (366 ft 9 in) *Beam:* 21.1 m (69 ft 2 in) *Draught:* 7.27 m (23 ft 10 in) *Machinery:* 2-shaft steam triple-expansion, 13500 ihp=18 knots *Protection:* 250 mm (9.8 in) belt; 70 mm (2.75 in) deck; 250 mm (9.8 in) turrets *Armament:* 4 10-in (254-mm); 8 6-in (152-mm); 8 4.7-in (120-mm); 8 57-mm (2.24-in); 2 37-mm; 4 17.7-in (45-cm) torpedo tubes (above water) *Crew:* 537

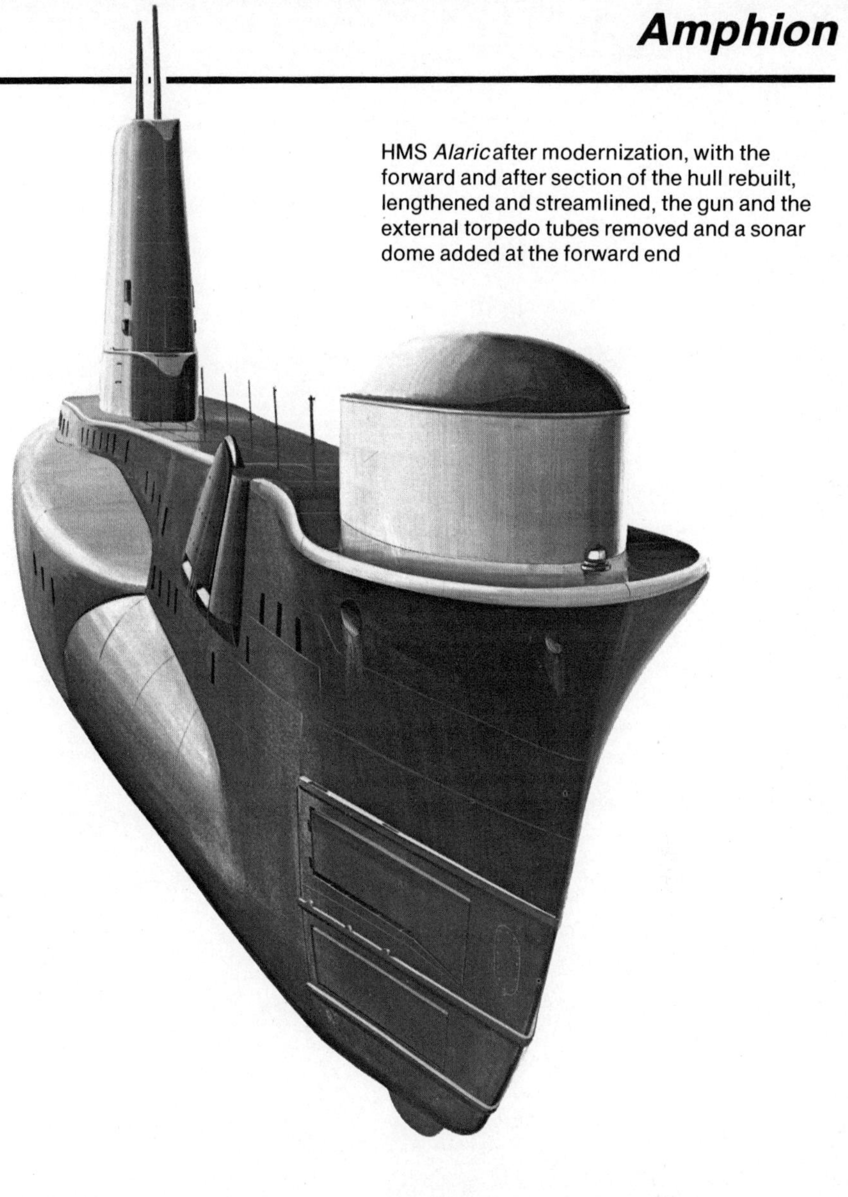

HMS *Alaric* after modernization, with the forward and after section of the hull rebuilt, lengthened and streamlined, the gun and the external torpedo tubes removed and a sonar dome added at the forward end

Amphion

British submarine class. Intended to operate in the Far East and Pacific, these submarines were specifically designed for mass-production using prefabrication and welding techniques. These methods had been used in earlier submarines but could not be employed to their full advantage because the boats in question had not originally been designed for this type of construction. The *Amphion* or 'A' Class were the first British submarines designed to have a completely welded hull. The system gave considerable advantages in that the prefabricated sections could be built under cover in ideal conditions before being transferred to the building slip for assembly. The slip was, moreover, occupied for the shortest possible time, which allowed a higher rate of production.

The 'A' Class pressure hull was constructed from 10 main sections and the ballast tanks from 16. Welding also improved hull strength, which enabled the boats to dive to a greater depth and saved weight. Compared to the earlier 'T' Class the 'A' Class incorporated several improvements without a major increase in size. They had a range of 16800 km (10500 miles) at 11 knots and a surface speed of 18.5 knots compared with 12800 km (8000 miles) at 10 knots and 15.25 knots for the 'T' Class. This high surface speed and radius of action were essential for their intended area of operations in the Pacific and Far East.

The gun armament was not always carried, being fitted or not according to operational requirements. Four of the ten torpedo tubes were fitted in the casing outside the pressure hull (two forward and two aft). The remainder were mounted in the pressure hull (four forward and two aft). To enable the diesels to be run while the boat was submerged they were fitted with a hinged 'snort', developed from the German *Schnorchel*, containing the engines' inlet and exhaust pipes. They were also fitted with two air-conditioning plants and an air-purification system, which war experience had demonstrated to be essential in areas of extreme climate such as the Far East, to keep the crew at peak efficiency.

Forty-six of the class were ordered between 1944 and 1945 but at the end of hostilities 30 were cancelled. Of these 30 only two, *Ace* and *Achates*, had been launched. The remaining 16 were completed between 1946 and 1948. Between 1955 and 1962 the class was modernized. The old casing, including the four external tubes, was removed and replaced by a modern streamlined casing with a sonar dome at the forward end. The conning tower was replaced by a tall streamlined 'dorsal fin' which enclosed the navigating position, snort, periscope standards, and radio and radar aerials (retracted), thereby eliminating the underwater resistance to propulsion caused by these items. This streamlining, together with the weight saved by the use of aluminium and the removal of the external torpedo tubes, provided a slight improvement in speed. No position was provided for the 20-mm gun but there was one for the 4-in gun, although this was seldom carried.

On April 16, 1951, the *Affray* disappeared while operating in the English Channel. A search was instituted and she was eventually found lying on the bottom of Hurd Deep, off Alderney, in 55 fathoms (100 m) of water. There were no external signs of damage except for a broken 'snort' tube (unlikely by itself to cause her loss). No attempt was made at salvage for reasons of cost and the great depth, and her loss remains a mystery. The remainder of the class were scrapped between 1967 and 1975, starting with *Aurochs*, the only unconverted of the class.

Alcide, Alderney, Alliance, Ambush, Amphion, Anchorite, Andrew, Astute, Auriga, Aurochs—built by Vickers Armstrong.
Aeneas, Alaric, Affray—built by Cammell Laird.
Artemis, Artful—built by Scotts.
Acheron—built by Chatham dockyard.

Displacement: 1120/1620 tons (surfaced/submerged) *Length:* 85.8 m (281 ft 9 in) *Beam:* 6.78 m (22 ft 3 in) *Draught:* 5.18 m (17 ft) *Machinery:* 2-shaft 8-cylinder diesels, 4300 bhp=18½ knots (surfaced); 2-shaft electric motors, 1250 shp=8 knots (submerged) *Armament:* 10 21-in (53-cm) torpedo tubes; 1 4-in (102-mm) gun; 1 20-mm AA *Crew:* 61

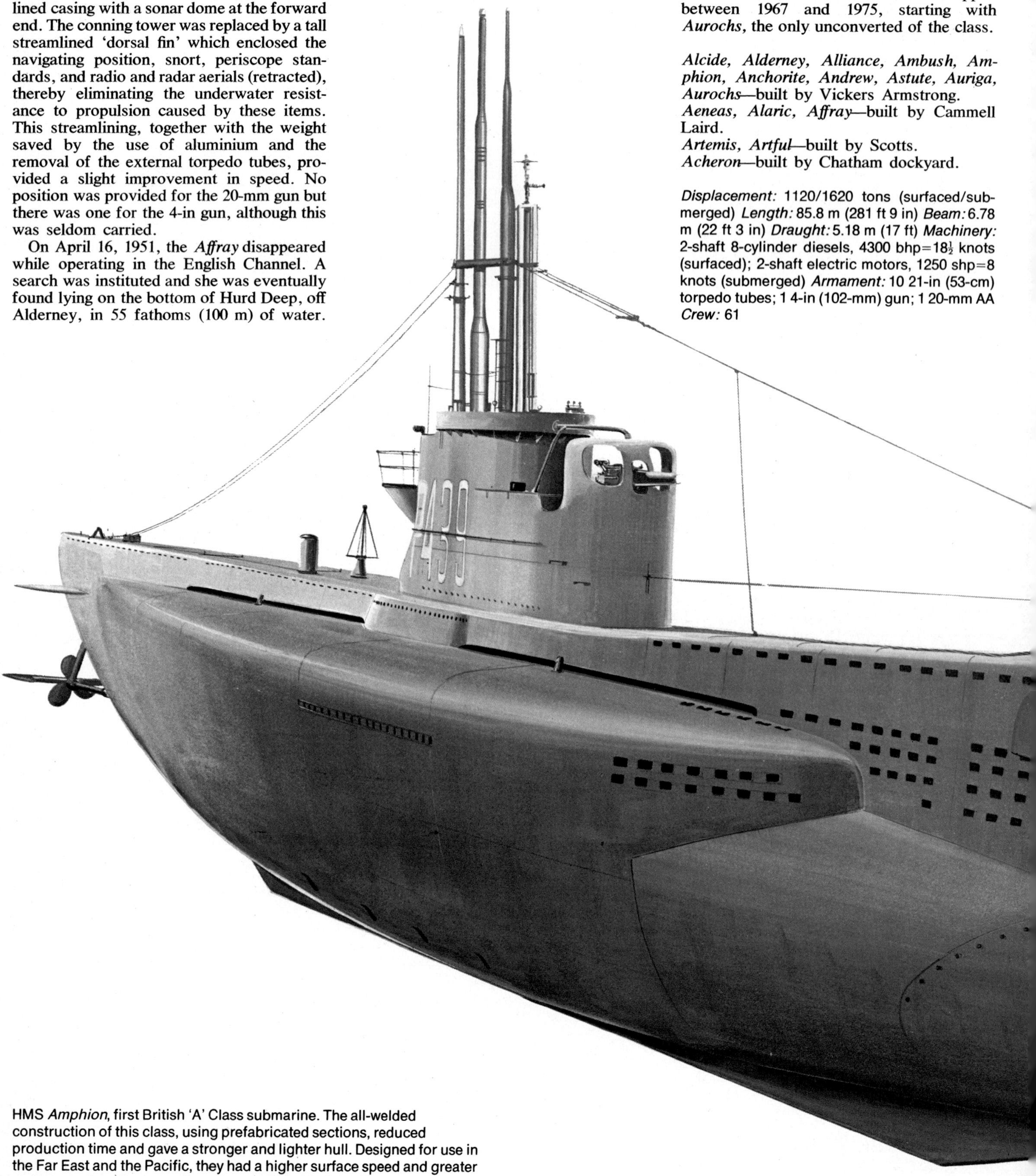

HMS *Amphion*, first British 'A' Class submarine. The all-welded construction of this class, using prefabricated sections, reduced production time and gave a stronger and lighter hull. Designed for use in the Far East and the Pacific, they had a higher surface speed and greater endurance than the preceding classes. Only 16 of the 46 boats ordered were completed, the remainder being cancelled in late 1945

Amphitrite

French submarine class. The ten vessels of this group were authorized under the 1912 Estimates and constructed between 1912 and 1914. Like the majority of pre-First World War French submarines they were of the *Laubeuf* type and were an improved and faster development of the *Brumaire* (or modified *Pluviôse*) Class. The *Clorinde* and *Cornélie* were slightly different from the others of the class. Two were converted to minelayers during the First World War.

They spent the majority of their wartime careers in the Mediterranean, one of the class being a war loss. The *Ariane* was torpedoed and sunk off the entrance to the Gulf of Bizerta by a German submarine on June 19, 1917. The remainder were sold for scrap during the 1920s.

Amarante, Aréthuse, Artémis, Atalante, built at Toulon dockyard.
Andromaque, Ariane built at Cherbourg dockyard.
Amphitrite, Astrée, Clorinde, Cornélie built at Rochefort dockyard.

Displacement: 413 tons (surface) 597 tons (submerged) *Length:* 53 m (174 ft) *Beam:* 5.02 m (16 ft 6 in) *Draught:* 3.04 m (10 ft) *Machinery:* 2-shaft diesel, 1300 hp=14 knots (surfaced); electric motors, 700 hp=8 knots (submerged) *Armament:* 1 3-pdr; 8 18-in (46-cm) torpedoes (two bow tubes external dropping gear). *Crew:* 30

Marius Bar

The French submarine *Amphitrite*, one of a class of ten built between 1912 and 1914

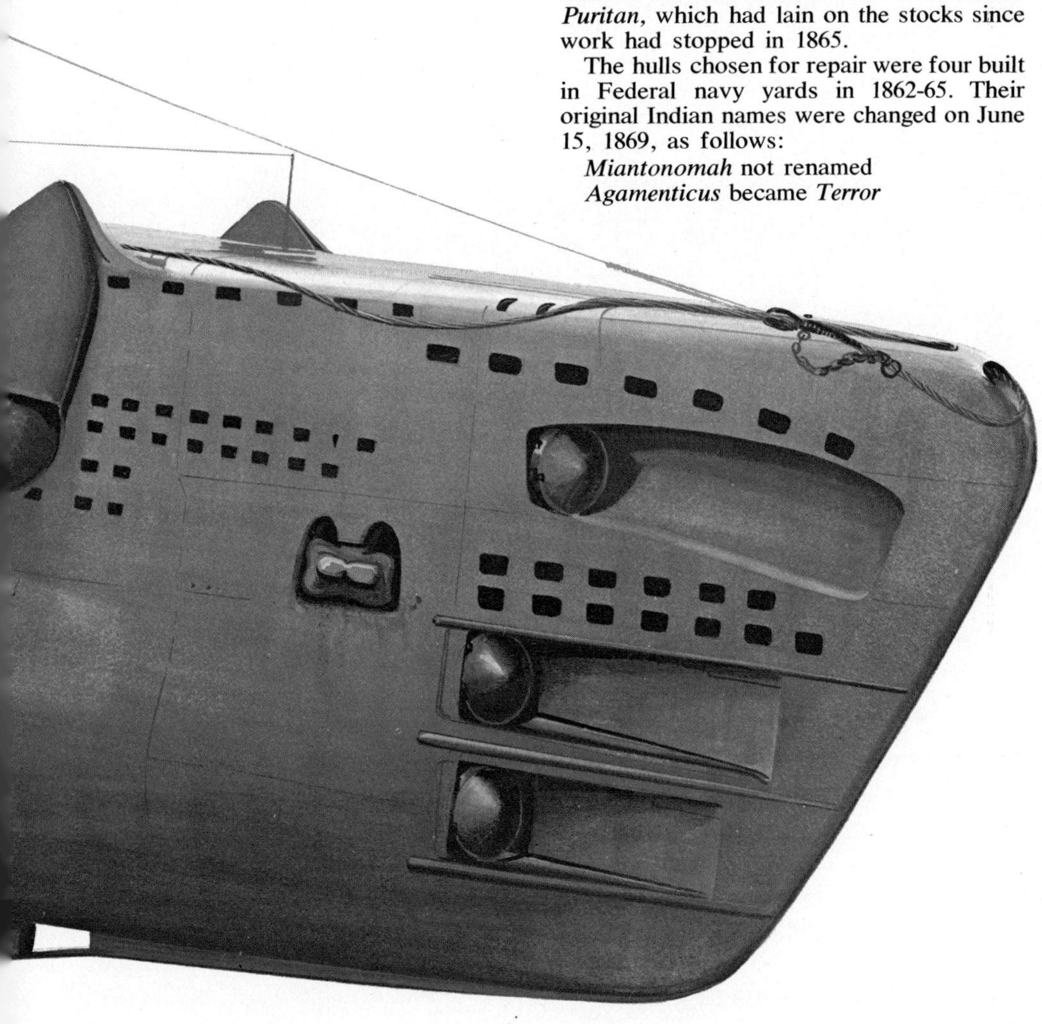

Amphitrite

US monitor class. After the Civil War of 1861-65 the US Navy laid up all its monitors, and as most of them had been built hurriedly with green timber they quickly developed dry rot. No new major warships were sanctioned by Congress between 1866 and 1874, and in desperation the Secretary of the Navy, George Robeson, obtained funds to repair four of the best monitors and to finish the *Puritan*, which had lain on the stocks since work had stopped in 1865.

The hulls chosen for repair were four built in Federal navy yards in 1862-65. Their original Indian names were changed on June 15, 1869, as follows:

Miantonomah not renamed
Agamenticus became *Terror*
Monadnock not renamed
Tonawanda renamed *Amphitrite*

Only the *Monadnock* saw any war service as she was completed in October 1864, but *Miantonomoh* crossed the Atlantic and *Monadnock* rounded Cape Horn. But by 1874 their fighting value had dwindled to nothing as they were so rotten, and Robeson decided to replace them entirely. To achieve this, when forbidden to build warships by a Congress dedicated to saving money on defence, he resorted to the 18th-century device of a 'great repair', and pretended that the new ships were actually the 1862 originals. However, the particulars show clearly that two different classes were involved and there is even a photograph of the old *Miantonomoh* ready for scrapping in 1874.

The new class, now known as the *Amphitrite* Class rather than the *Miantonomoh* Class, was ordered from private shipbuilders but finished by the Navy. Construction was lengthy, partly because the Navy Department wanted to avoid a suspiciously large expenditure of money.

Amphitrite: launched 1883 by Harlan & Hollingsworth, Wilmington, Del., and commissioned 1895 at Norfolk Navy Yard, Va.

Miantonomoh: launched 1876 by John Roach & Son, Chester, Pa., and commisioned 1891 at New York Navy Yard.

Monadnock: launched 1883 by Continental Ironworks, Vallejo, Calif., and commissioned 1896 at Mare Island Navy Yard, San Francisco.

Terror: launched 1883 by Wm. Cramp & Sons, Philadelphia, Pa., and commissioned 1896 at New York Navy Yard.

In 1898 the *Monadnock* crossed the Pacific to serve in the war against Spain but the monitor's low freeboard and lack of speed made them useless for anything but coast defence. During the First World War they gained a new lease of life as submarine tenders, a role for which their low freeboard was ideal. *Terror* and *Miantonomoh* became targets in 1915, and were sold with their sisters in 1920-23. Prior to being stricken they received hull numbers under the US Navy's reclassification scheme: BM 2 *Amphitrite*, BM 3 *Monadnock*, BM 4 *Terror*, BM 5 *Miantonomoh.*

Amphitrite

The *Amphitrite* was sold in 1920 and converted to a floating hotel. As such she survived the Second World War.

(Original) *Displacement:* 3400 tons (normal) *Length:* 78.8 m (258 ft 6 in) (oa) *Beam:* 16.1 m (52 ft 9 in) *Draught:* 3.85 m (12 ft 8 in) *Machinery:* 2-shaft horizontal reciprocating (ihp not known)=9 knots *Protection:* 5 in (127 mm) side, 10 in (245 mm) turrets *Armament:* 4 15-in (381-mm) Dahlgren smooth-bore muzzle-loaders (2×2) *Crew:* 150

(As rebuilt) *Displacement:* 3990 tons (normal) *Length:* 80.1 m (262 ft 9 in) (oa, average) *Beam:* 17 m (55 ft 10 in) (average) *Draught:* 4.42 m (14 ft 6 in) (mean) *Machinery:* 2-shaft horizontal compound reciprocating, 1600 ihp=12 knots (*Monadnock* had triple-expansion engines, 3000 ihp=$14\frac{1}{2}$ knots) *Protection:* 7-9 in (177-228 mm) side; $7\frac{1}{2}$-$11\frac{1}{2}$ in (190-292 mm) turrets (varying from ship to ship) *Armament:* 4 10-in (254-mm) breechloaders; 2 4-in (102-mm) (*Miantonomoh* no 4-in, *Terror* 4 4-in); 2 6-pdrs (57-mm) *Crew:* 182

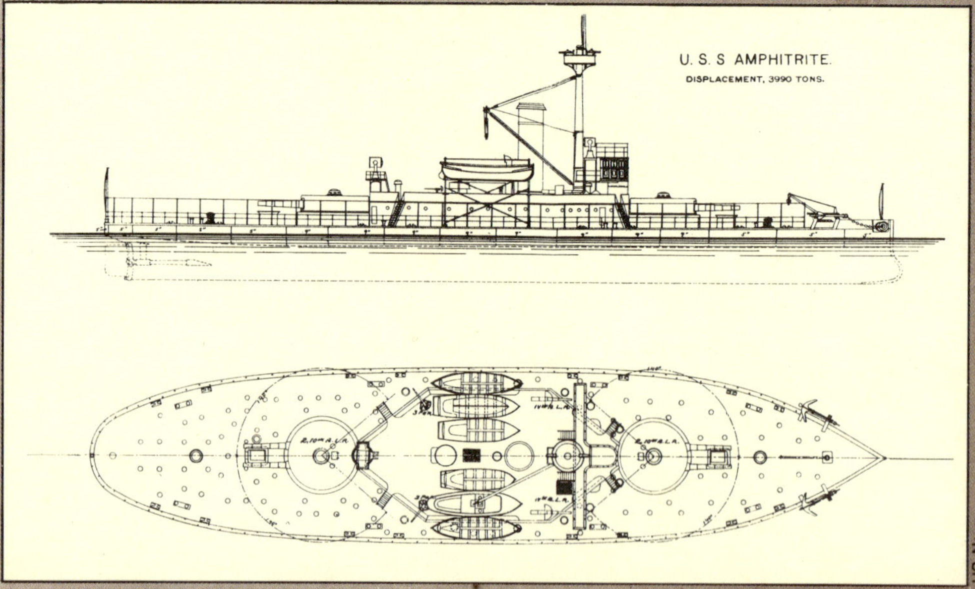

US Navy

An unofficial plan of the monitor USS *Amphitrite* which was published in 1893 in the transactions of the Society of Naval Architects and Marine Engineers

USS *Monadnock*, third of the *Amphitrite* Class of monitors, on her way from San Francisco to Manila in 1898 to add her four 10-in guns to the war against the Spanish. The ships were supposedly repaired Civil War vessels: in fact, they were completely new warships, the fiction of the 'great repair' and the use of the names of the older ships being a ruse to enable new ships to be built against the will of Congress. Their poor sea-going qualities and extremely low freeboard made them useless for anything but coast defence, though the latter characteristic made them ideal submarine tenders, a role they filled during the First World War

US Navy

AMST

US Air Force
transport aircraft programme.

Though Lockheed-Georgia can claim that the best replacement for the C-130 Hercules airlifter is a later C-130, the US Air Force decided in early 1972 to fund a competition to find an 'Advanced Medium STOL Transport' (AMST), with a view to eventual takeover from the evergreen 'Herky-bird'. From five submissions the USAF selected radically different proposals by Boeing (the YC-14) and McDonnell Douglas (YC-15). Each was awarded a contract for construction and flight test of two prototypes, with the intention of putting the aircraft judged superior into production as the main tactical airlifter of the 1980s.

McDonnell Douglas's contract, at $85.9 million, has produced an aircraft considerably bigger in volume capacity than the C-130, with the ability to carry 12 247 kg (27 000 lb) cargo in STOL operation from a rough 610-m (2000-ft) strip, or 28 122 kg (62 000 lb) from a regular runway. The four 7257-kg (16 000-lb) thrust Pratt & Whitney JT8D-17 turbofans are podded well forward of the wing but discharge immediately below the undersurface, so that when the wide-span, multi-section flaps are lowered the whole engine efflux, plus entrained air, is deflected sharply down to give greatly increased lift for STOL. The YC-15 has an advanced flight-control system, with front and rear rudder sections and powerful upper-surface wing spoilers used as a direct-lift control. McDonnell Douglas will test-fly two new engines, the refanned JT8D-209 and CFM56 'ten-ton' turbofan, in YC-15s during 1977. The two YC-15 aircraft flew in August and December 1975, and completed initial test flying in August 1976.

Boeing's contract, priced at $105.9 million, has resulted in a more difficult but more advanced machine using the radical upper-surface blowing technique. Two General Electric F103 (CF6-50D) turbofans, each rated at 23 133 kg (51 000 lb) thrust, are mounted ahead of the small supercritical wing, blowing across the upper surface. This can also generate greatly augmented STOL lift, with a clear underside for carrying RPVs or other stores, with minimal noise and infrared signature and more efficient thrust reversal. Financial hold-ups delayed the first YC-14 about a year, but Boeing refined and

Boeing's competitor for the AMST contract, the YC-14. Upper surface blowing from the two turbofan engines gives augmented STOL lift and clear wing undersides for stores points

Boeing Aerospace

Takeoff of a YC-15. Upper surface spoilers on the wings are used as direct-lift controls and the engine efflux immediately below the wings gives enhanced lift for STOL operations

McDonnell Douglas

The McDonnell-Douglas YC-15 AMST prototype gives a LAPES (Low-Angle Parachute Extraction System) demonstration

Twenty-seven tons of a US Army 155-mm SP howitzer are loaded into the YC-15, whose specification calls for the ability to lift a maximum of 28 123 kg (62 000 lb), or 12 247 kg (27 000 lb) in STOL operations

McDonnell Douglas

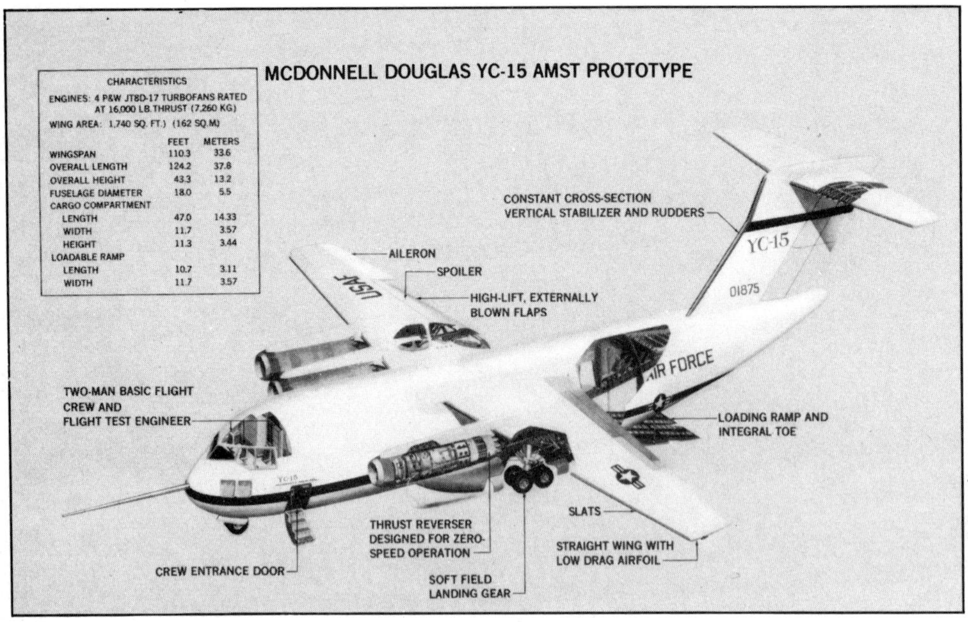

McDonnell Douglas

improved the design as a result and judge it to be superior to its slightly smaller rival. While restricted to the same STOL payload as the YC-15, the YC-14 can lift 36 740 kg (81 000 lb) in conventional runway missions. Among its extremely advanced features are a mainly British electrically signalled triple-redundant flight-control system.

(Boeing YC-14) *Span:* 39.32 m (129 ft 0 in) *Length:* 40.13 m (131 ft 8 in) *Gross weight:* 112 945 kg (249 000 lb) *Max speed:* 834 km/h (518 mph)

(McDonnell Douglas YC-15) *Span:* 33.63 m (110 ft 4 in) *Length:* 37.8 m (124 ft 0 in) *Gross weight:* 99 335 kg (219 000 lb) *Max speed:* 861 km/h (535 mph)

Amur and *Yenisei*

Russian minelayers. These vessels can easily lead one into a state of confusion as two pairs were built with the same names, at the same yard and to much the same design. The first pair were laid down at the Baltic Works, St Petersburg, in 1898, the *Amur* being completed in 1899 and the *Yenisei* in 1900. They were the first oceangoing minelayers and could lay mines while at full speed.

On the outbreak of the Russo-Japanese war in 1904 both vessels were in the Far East at Port Arthur. On February 11, 1904, the *Yenisei* was sunk by one of her own mines while laying a defensive minefield in Talien Bay. On May 15, 1904, the Japanese battleships *Hatsuse* and *Yashima* were mined and sunk 19 km (12 miles) off Port Arthur in a field laid by the *Amur* on the previous day. The *Amur* was eventually wrecked by Japanese land artillery while in dry dock at Port Arthur. She was salvaged by the Japanese in 1905 and scrapped.

The second pair were laid down at the Baltic yard in 1905, the *Amur* being completed in 1909 and the *Yenisei* in 1910. They were built to the same basic design as the first pair but were made slightly larger in order to give them an increased mine capacity and gun armament. Both served in the Baltic during the First World War, the *Yenisei* being torpedoed and sunk by the German submarine, *U 26*, off Odenstrolm on June 4, 1915. *Amur*'s gun armament was altered to nine 4.7-in and one 3-in AA during the war. After the revolution the *Amur* served in the Soviet Fleet as a training ship until 1940 when she was hulked at Tallinn. Here she was employed as a mine depot ship until scuttled to avoid capture on August 28, 1941. She was later refloated and remained in service until scrapped in the early 1950s.

(First Pair) *Displacement:* 2590 tons *Length:* 92.04 m (302 ft) *Beam:* 12.19 m (40 ft) *Draught:* 5.99 m (19 ft 8 in) *Machinery:* 2-shaft vertical triple-expansion steam engines, 5200 ihp=17 knots *Armament:* 5 3-in (75-mm) (5×1); 5 6-pdr (5×1); 300 mines

(Second Pair) *Displacement:* 3020 tons *Length:* 97.84 m (321 ft 6 in) *Beam:* 14.02 m (46 ft) *Draught:* 4.48 m (14 ft 9 in) *Machinery:* As above *Armament:* 5 4.7-in (120-mm) (5×1); 2 3-in (77-mm) (2×1); 360 mines

Amuric

Russian destroyer class. Two of this class, the *Gaidamak* and *Vsadnik*, were constructed in the Krupp-Germania yard at Kiel between 1904 and 1906. The remaining two, *Amuric* and *Ussurietz*, were built at Riga and Helsingfors respectively between 1904 and 1907 with materials supplied by Krupps. Being of German design they closely resembled contemporary German destroyers. The torpedo tubes were all mounted on the centre line, being positioned between the two funnels, abaft the second funnel and abaft the mainmast. The two guns were mounted in the usual fore and aft positions. Like the majority of Russian destroyers they were also equipped to carry mines. All four served in the Baltic Fleet and after the Revolution became part of the Soviet Fleet. Three were renamed, the *Amuric* became *Zeleznja*, the *Vsadnik* became *Sladkov* and the *Ussurietz* became *Rosal*.

Displacement: 570 tons *Length:* 70.1 m (233 ft) *Beam:* 7.32 m (24 ft) *Draught:* 2.28 m (7 ft 6 in) *Machinery:* 2-shaft triple expansion steam engines, 6500 ihp=25 knots *Armament:* 2 4-in (102-mm) (2×1); 1 1-pdr; 3 18-in (46-cm) torpedo tubes (3×1); 25 mines *Crew:* 98

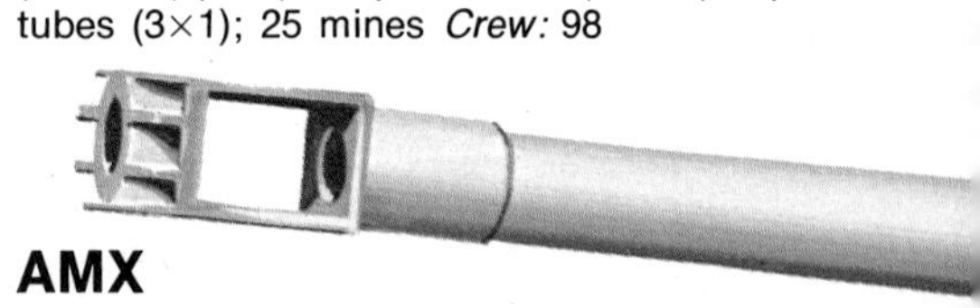

AMX

Atelier des Constructions d'Issy-les-Moulineaux (AMX) is the French army's design and development facility for armoured vehicles. This establishment began the study of tank design as the Second World War was ending, their first product being the AMX-50—the number, as in all subsequent designs, indicating the stipulated weight in metric tonnes. This was to be a universal battle tank with high mobility, strong protection and powerful armament. Armed with a 90-mm gun, then a 100-mm and finally a 120-mm, it

ECP Armées

AMX-13 with 90-mm gun, showing the turret tilted back to elevate the gun

was powered by a 1000-bhp fuel injection engine based on the best features of German wartime design.

Although development continued for several years, the AMX-50 was not accepted for service, but it pointed the way to the feasibility of a single type of tank and rendered obsolete the old distinctions between 'cruiser', 'infantry', 'support' and other types of tank.

The AMX-50 was turned down largely because the French army were convinced that the advantages of massive tanks were less than their disadvantages, particularly as by the early 1950s it had been amply demonstrated that contemporary antitank weapons could defeat any thickness of armour capable of being carried on a practical vehicle. A fresh design was begun, the AMX-38, but while this was in the prototype stage the French, German and Italian governments, in 1957, agreed to develop a tank to a tripartite formula. Like several such ventures this came to nothing and the partners each went their own ways, Germany to the Leopard, Italy to adopt the US M60 tank, and France to evolve the AMX-30, and the AMX-38 project was abandoned.

The AMX-30 entered French army service in 1967 as their main battle tank. Armed with a 105-mm gun, the standard projectile of which is a non-rotating hollow charge shell of exceptional efficiency, the tank weighs 36 tonnes, travels at 64 km/h (40 mph) and gives a good balance between mobility, firepower and protection. The basic chassis has since been adapted to produce a recovery vehicle, a bridge-carrying tank, a tracked launch vehicle for the Pluton and Roland missiles and an antiaircraft tank with twin 30-mm guns.

While the AMX-50 was under development, in the late 1940s, the French army expressed interest in a tank capable of being air-transported, and in 1949 the first prototype of the AMX-13 was announced. Although the weight came out at 14.3 tonnes it could still be carried by the Breguet Deux-Ponts' transport aircraft, but in the event only four of the military transports were produced and the airborne aspect of the AMX-13 faded away.

RAC Tank Museum

The AMX-50t, first of the AMX tank series, in 1951, with the original 90-mm gun

The original AMX-13, with its 75-mm gun mounted rigidly in a trunnioned turret. The whole turret is tilted to elevate the gun, and the rigid mounting enables an automatic loading gear to be used

AMX

Instead it became a light tank and tank destroyer. It was fitted with a high velocity 75-mm gun rigidly mounted in a unique turret. The whole turret was trunnioned so as to tilt in order to elevate the gun. As a result, the gun could be rigidly mounted (except for the recoil movement) and an automatic loading gear was built into the turret, thus dispensing with the need for a human loader and reducing the crew to three men. The AMX-13 was produced in large numbers for the French and later sold to 23 other countries, including Israel, by whom the tank was used in the 1956 campaign in Sinai.

The AMX-13 chassis has formed the basis for several specialist vehicles: the AMX-13-51 which carries four SS.11 missiles in addition to the normal gun; another with six Hot missiles; an armoured recovery tank and a bridging tank. The basic tank has also been upgunned at various times; the original FL-10 turret with long 75-mm gun was supplemented by tanks using the FL-11 turret with short 105-mm support gun in Algeria, and more recently a 90-mm gun version has been produced. It was then decided to retrofit the original 75-mm gun tanks with the new 90-mm weapon.

A lengthened and widened version of the AMX-13 chassis formed the starting point for a family of armoured personnel carriers and support vehicles introduced in the 1950s. Known variously as the AMX-VTP (Vehicule Transport de Personnel), the TT CH Mle 56 (Transport de Troupe Chenille Modèle 1956) or the AMX-VCI (Vehicule de Combat d'Infanterie), the APC carries a crew of 13 men and is armed with a heavy machine-gun in a turret or ring mount. The occupants sit facing outwards and can fire their personal weapons through ports in the vehicle sides. The remainder of the family consists of a command vehicle, an ambulance, a dozer tractor, an artillery command post, a mortar carrier with either 81-mm or 120-mm mortar, a pioneer vehicle with dozer blade, winches, lifting gear and similar equipment, a cargo carrier, a missile launcher carrying Entac missiles, and an artillery ammunition carrier.

Under development in the mid-1970s was a complete new series of armoured tracked vehicles known as the AMX-10 series. These are an improvement on the earlier vehicles, of much cleaner design and better performance, and they are specifically intended as fighting vehicles rather than simply carriers. The basic vehicle is the AMX-10P which weighs 13.8 tonnes in combat order and is armed with a 7.62-mm machine-gun and a 20-mm cannon capable of firing at 700 rds/min. Both weapons are coaxially mounted above the cupola and are fired by remote control from inside the vehicle. It is also proposed to fit a Milan antitank missile launcher. The vehicle has an operating crew of two—driver and gunner—and carries a nine-man infantry section, the section commander acting as vehicle commander.

Further versions under development include the AMX-10C armed with a 105-mm howitzer in a turret; the AMX-10M, one version of which carries a 90-mm gun in the hull front to fire the Acra missile, while another proposed version carries four Hot missiles; the AMX-10D recovery vehicle and the AMX-10PC command vehicle. There is also a development programme for a six-wheeled armoured car version of the design, known as the AMX-10R, and three models, the RP, RC and RM, corresponding to the P, C and M tracked models, are being evaluated. The basic 10P has been adopted by the French army, and numbers of the variant models are expected to be accepted.

The AMX-13 chassis also formed the basis for two self-propelled 105-mm howitzer designs. The Model A carries the howitzer in a fixed superstructure, while the Model B carries it in a rotating turret. Both models allow the howitzer to elevate to 70°, while the Model A allows 20° of traverse each side of the centre and the Model B has all-round traverse. Maximum range of the howitzer is 15 000 m (16 400 yds). Both models entered service with the French army, while numbers of the Model A have also been bought by the Netherlands and Israel. See also APC.

ECP Armées

Above: **The AMX-30 main battle tank entered French service in 1967 with a 105-mm gun.** ***Below:*** **AMX-10P light AFV with 20-mm cannon and 7.62-mm machine-gun**

ECP Armées

AMX COMPARATIVE DATA

Feature	AMX-30	AMX-13	AMX-VTP	AMX-10P
Weight (combat loaded)	36 tons	14.8 tons	14 tons	13.8 tons
Length overall	31 ft 2 in/9.5 m	20 ft 10 in/6.35 m	18 ft 4 in/5.6 m	19 ft 2 in/5.84 m
Width	10 ft 2 in/3.1 m	8 ft 3 in/2.5 m	8 ft 7 in/2.6 m	9 ft 1 in/2.76 m
Speed (firm surface)	40 mph/64 km/h	37 mph/59 km/h	37 mph/59 km/h	40 mph/64 km/h
Range	300 miles/482 km	210 miles/338 km	210 miles/338 km	373 miles/595 km
Armament	105-mm	75-mm	12.7-mm	20-mm
Armour	na	10-40 mm	10-30 mm	na
Engine bph	720	260	250	276

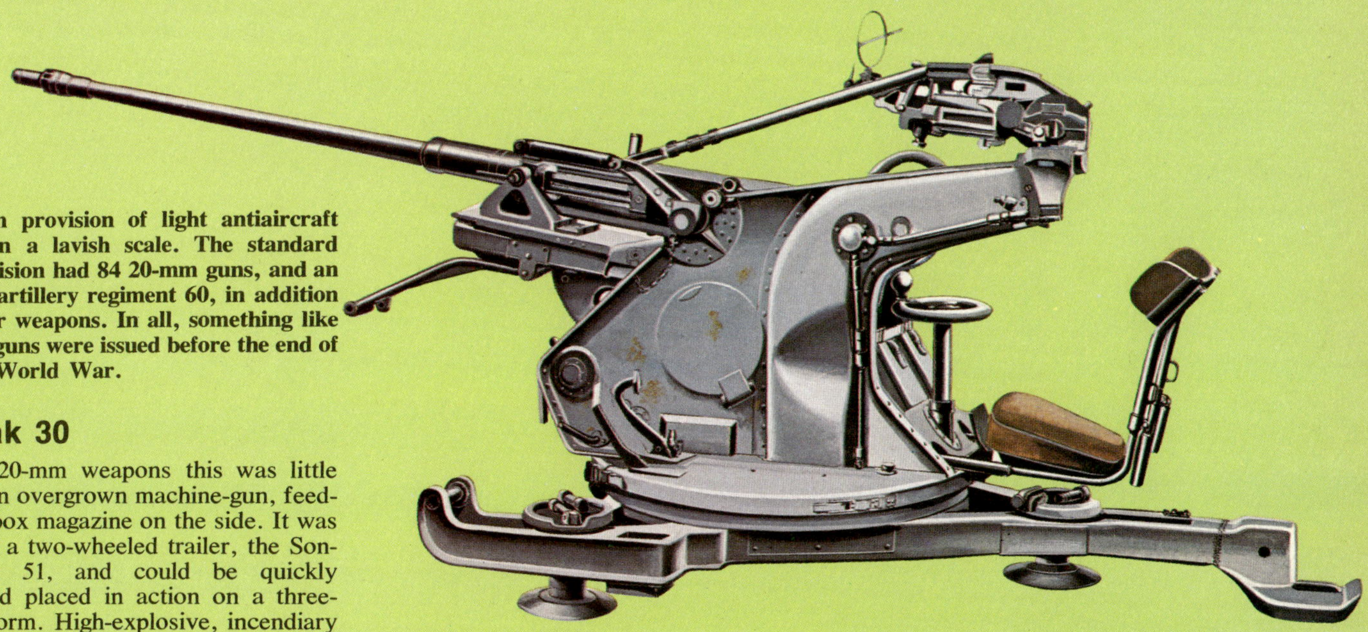

The 2-cm Flak 30 light antiaircraft gun in the operational position on its three-legged platform. The gun was normally transported on its two-wheeled Sonderanhanger 51 trailer; able to fire all the basic types of ammunition it could also double as a light field gun

The German provision of light antiaircraft guns was on a lavish scale. The standard Infantry Division had 84 20-mm guns, and an antiaircraft artillery regiment 60, in addition to its heavier weapons. In all, something like 30 000 light guns were issued before the end of the Second World War.

2-cm Flak 30

Like most 20-mm weapons this was little more than an overgrown machine-gun, feeding from a box magazine on the side. It was mounted on a two-wheeled trailer, the Sonderanhanger 51, and could be quickly removed and placed in action on a three-legged platform. High-explosive, incendiary and armour-piercing ammunition were provided, and the gun could double as a light antitank weapon if necessary, though this role was soon overtaken by improvements in tank armour.

I V Hogg

The 2-cm Flak 30 on its two-wheeled trailer

2-cm Flak 38

The Flak 30 served in the 1939 Polish campaign, and this showed that a better rate of fire was wanted. The gun was redesigned by Mauserwerke, the changes being all internal and the improved weapon looking no different from the earlier model, and using the same two-wheeled carriage. In 1940 the Mauser company developed a four-barrelled model for the German navy and this, on a two-wheeled trailer of stronger construction, was taken into use as the *Flakvierling* by both the army and the Luftwaffe. Numbers were also fitted to semi-track tractors.

3-cm Flak 103/38

This was developed in 1944 in order to provide an increase in the number of light guns, and was a Rheinmetall aircraft cannon fitted to the basic 2-cm Flak trailer and mounting. Using a heavier shell and with a higher rate of fire it promised good performance, but the gun was too powerful for the mounting and the whole equipment was never very reliable. It was also used in two-barrel and four-barrel form on some self-propelled AA equipments. See 'Kugelblitz' and 'Zerstorer-45'.

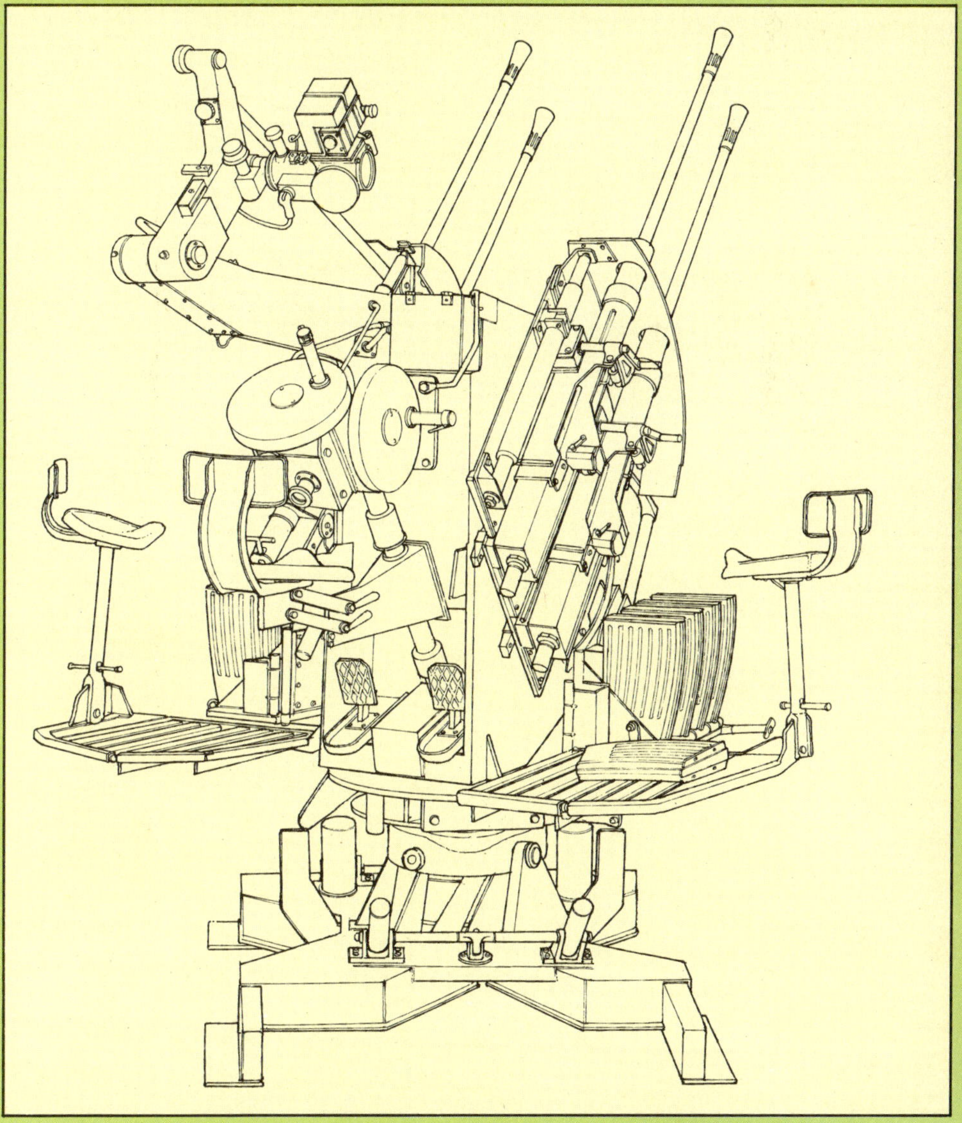

2-cm Flakvierling, the four-barrelled version of the Flak 38, used both on platforms and mounted on semi-tracked trucks for the protection of mobile columns

Light Flak

3.7-cm Flak 18

Introduced in 1935, this was little more than an enlarged version of the 2-cm Flak 30. It was not well received, the army reporting that it was too heavy and cumbersome for its calibre. Only a small number were made, but they remained in service for most of the war.

3.7-cm Flak 36 & 37

This was the redesigned Flak 18, and the changes were mainly in the mounting, using a three-legged platform instead of the original four-legged model, and a lighter and more compact two-wheeled trailer to carry it. The models 36 and 37 were identical except for the type of sight fitted. After the war a number of Flak 37 remained in service with the Rumanian army for some years.

3.7-cm Flak 43

This replaced the Flak 37 and was little more than an enlarged Rheinmetall 108 MK103 aircraft cannon. The same three-legged mount and two-wheeled trailer were used, suitably strengthened, and the most notable feature was the ammunition feed through the enlarged trunnions. This ensured that loading a full clip of shells did not upset the centre of gravity and gun balance and so alter the point of aim during firing. A two-barrelled version, the Flakzwilling 43, was also produced.

I V Hogg

The 3.7-cm Flakzwilling 43. The ammunition clips were fed to the gun via enlarged trunnions which kept the centre of gravity stable

I V Hogg

I V Hogg

The 3.7-cm Flak 18 mounted on a SdKfz 6/2 semi-track tractor. Many combinations of half-tracks and light antiaircraft pieces were developed during the war: this is one of the more common types. *Inset:* Flak 18 with its crew in Poland, 1939

COMPARATIVE DATA

Gun	Weight in Action kg/lb	Maximum Elevation degrees	Weight of Shell gm/oz	Rate of Fire rds/min	Muzzle Velocity m/sec/ft/sec	Maximum Horiz. Range m/yd	Maximum Vertical Range m/ft
2-cm Flak 30	483/1065	90	120/4.2	280	899/2950	2697/2950	2134/7000
2-cm Flak 38	406/895	90	120/4.2	450	899/2950	2697/2950	2134/7000
2-cm Flakvierling	15219/3352	100	120/4.2	1800	899/2950	2697/2950	2134/7000
3-cm Flak 103	618/1363	80	150/5.3	400	899/2950	5715/6250	4694/15400
3.7 Flak 18	1757/3858	85	556/19.6	160	820/2690	6492/7100	4785/15700
3.7 Flak 36	1544/3405	85	556/19.6	160	820/2690	6492/7100	4785/15700
3.7 Flak 43	1247/2750	90	556/19.6	250	820/2690	6584/7200	4785/15700
3.7 Flakzwilling	2781/6130	90	556/19.6	500	820/2690	6584/7200	4785/15700

The idea of tapering the barrel of a gun in order to obtain an increase in velocity attracted various inventors before it was successfully applied by Hermann Gerlich, a German weapons designer, in the late 1920s. Gerlich died in 1934 but he had pointed the way, and the German army took advantage of his designs and theories to produce three antitank guns. The ballistic principle is simple: if the base area of the shot diminishes as it passes down a tapering barrel, then the gas pressure per unit of base area increases and the shot is accelerated. Translating this idea into a practical gun was, however, a rather more difficult matter.

The three weapons all used tungsten-cored solid shot with a sheath of mild steel which was formed into skirts of the major calibre, so that these skirts were squeezed down during the travel through the barrel. Penetration was done by the tungsten and damage to the tank was by splintering. High-explosive shells were also developed, though as might be imagined this was a much more difficult design, and they were rarely used in action.

The guns were extremely efficient and successful, but the critical shortage of tungsten in Germany after 1942 put an end to the special ammunition and consequently to the guns themselves.

2.8-cm schwere Panzerbuchse 41

Although called a 'heavy antitank rifle' for secrecy's sake, this was a light gun, using a sliding breechblock, a recoil system, and a two-wheeled split-trail carriage. The calibre was 2.8 cm at the breech end, tapering evenly along the 1.35-m (53-in) barrel to 20 mm at the muzzle. The gun was freely mounted on the carriage and was aimed by the gunner grasping a pair of spade grips and swinging the barrel. There was also an airborne model, using a lightweight tubular steel carriage, produced for airborne forces.

I V Hogg

A captured schwere Panzerbuchse 41 being tested at Pendine in South Wales during 1941. The trials revealed a muzzle velocity of 1400 m/sec

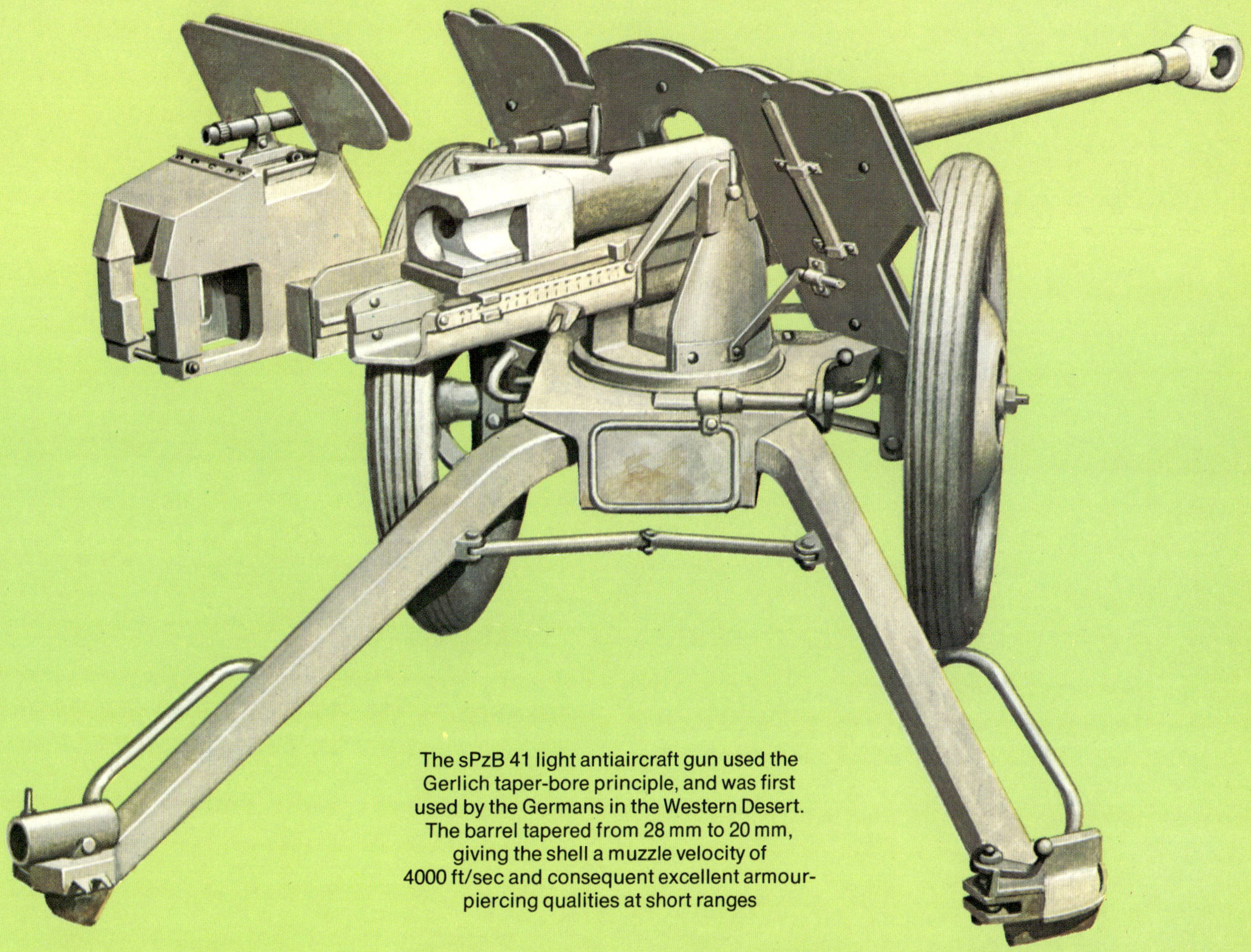

The sPzB 41 light antiaircraft gun used the Gerlich taper-bore principle, and was first used by the Germans in the Western Desert. The barrel tapered from 28 mm to 20 mm, giving the shell a muzzle velocity of 4000 ft/sec and consequent excellent armour-piercing qualities at short ranges

Taper-bore Antitank

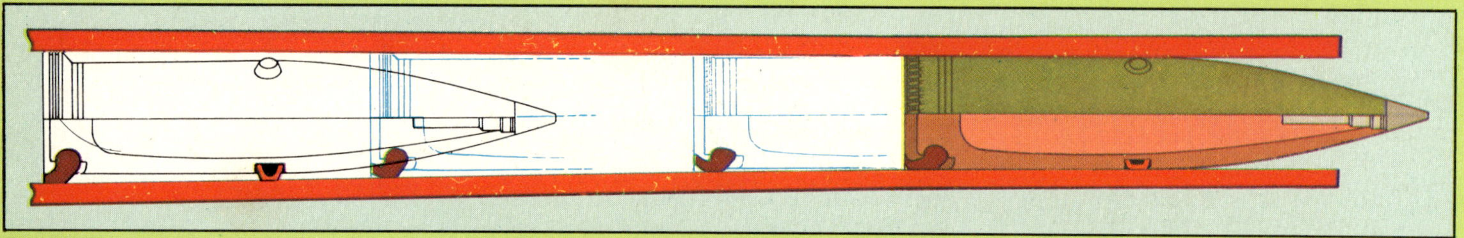

The Gerlich squeeze principle involves a barrel tapering towards the muzzle: as the shell travels along the barrel, the supporting studs are forced into their recesses and the sealing band is compressed, imparting a much higher muzzle velocity to the shot

COMPARATIVE DATA

Gun	Weight in Action kg/lb	Barrel Length cm/in	Weight of shot gm/oz	Muzzle Velocity m/sec (ft/sec)	Penetration at:	
					500m (550 yds)	1000m (1010 yds)
2.8 cm sPzB 41	229/505*	180/67.48	131/4.62	1402/4593	66 mm (2.6 in)	—
4.2 cm PzJK 41	642/1416	250/92.52	336/11.85	1267/4150	87 mm (3.4 in)	60 mm (2.4 in)
7.5 cm PAK 41	1356/2990	430/170.08	2.5 kg/5.71 lb	1127/3691	209 mm (8.2 in)	177 mm (7 in)

NOTES: *Airborne version weighed 118 kg (260 lb). Maximum range is not quoted, since this is never a criterion for antitank guns

4.2-cm Panzerjägerkanone 41

The second taper-bore gun, this had a calibre of 40.6 mm at the breech end, tapering to 29.4 mm at the muzzle. It was mounted on the same carriage as the conventional 3.7-cm PAK 36, and the two are extremely difficult to tell apart, the only obvious change being the length of the barrel, the 4.2-cm gun being about 76 cm (30 in) longer. It saw limited use in Italy and wider use on the Eastern Front, but production ended in 1942 due to a shortage of manganese (used in the carriage adapters for the new barrel) and tungsten (used in the ammunition).

7.5-cm Panzerabwehrkanone 41

This gun was made by Krupp in answer to the German army's 1939 demand for a heavy antitank gun. It differs from the other two in that the barrel does not taper evenly from breech to muzzle but is a parallel-sided barrel with a smooth-bored 'squeeze' adapter added at the front end. The barrel is thus parallel for 2.95 m (116 in), then tapers at 1 in 20 for 25 cm (10 in), then more steeply at 1 in 12 for 18 cm (7 in), finally becoming parallel for the last 61 cm (24 in).The two squeeze sections reduce the projectile from 75-mm calibre to 55-mm calibre and the designers claimed that this system of manufacture had benefits in the design of the ammunition and also allowed the squeeze section to be removed and replaced in the field when it became worn. The carriage was also of novel design, a split-trail two-wheeled type in which the shield was part of the basic structure and carried the barrel in a ball-mounting. Like the other taper-bore guns, the PAK 41 was made redundant when the supply of special ammunition ran out in 1942, but some of the carriages were refitted with the barrel of the conventional 7.5-cm PAK 40 gun.

I V Hogg

Production of the 4.2-cm PzJK was terminated in 1942 because of the tungsten shortage

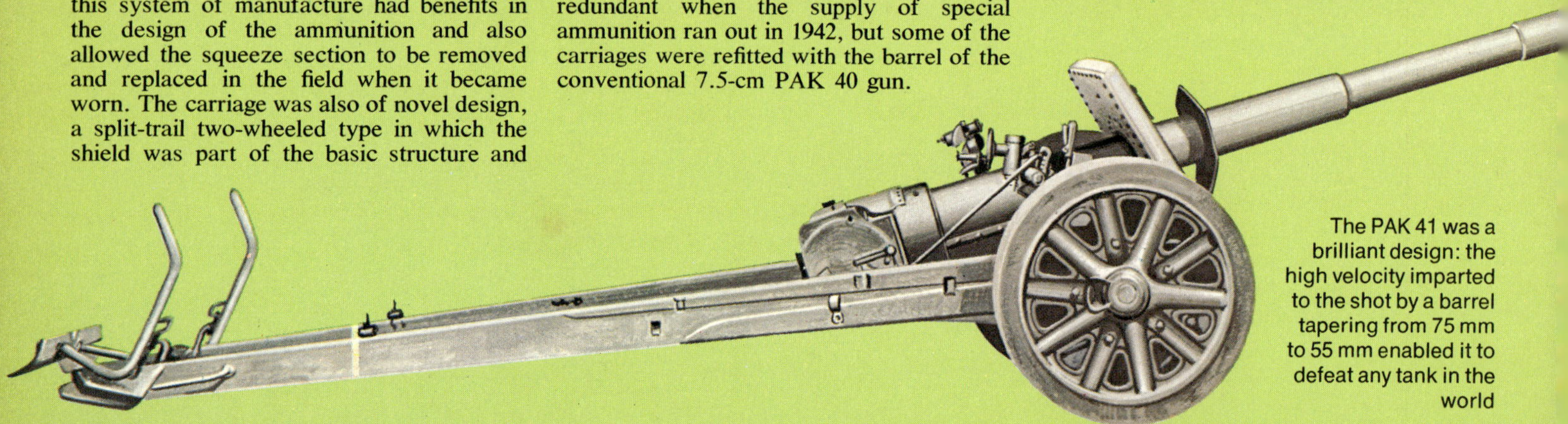

The PAK 41 was a brilliant design: the high velocity imparted to the shot by a barrel tapering from 75 mm to 55 mm enabled it to defeat any tank in the world

I V Hogg

Right: **The 7.5-cm Geb K 15, the German army's first mountain gun.** ***Above:*** **The Geb K 15 in action with its shield removed**

Although mountain guns are a highly specialized form of artillery—needing to be light yet stable, easy to dismantle into pack loads, and have a considerable range of depression and elevation to cope with the extremes of mountain fighting—and some peculiar weapons have been produced for mountain troops, the German equipments were extremely well-designed and workmanlike.

7.5-cm Gebirgskanone 15

The 7.5-cm Geb K 15 was the first mountain gun to be provided. This was a Skoda design adopted simply as a stopgap until better, German-designed, weapons could be produced. Firing a 5.1-kg (12.06-lb) shell it could reach a maximum range of 662.4 m (7245 yards) and could elevate from −9° to +50°. It could be stripped into seven pack loads, the heaviest of which weighed 151.5 kg (344 lb).

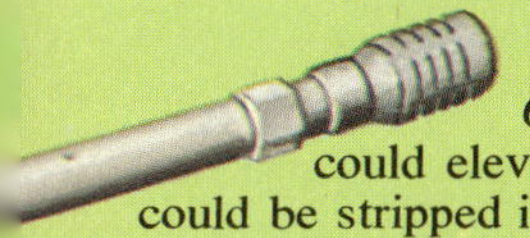

7.5-cm Gebirgsgeschütz 36

After 1938 this gun, designed by Rheinmetall-Borsig, replaced the 7.5-cm Geb K 15. A more up-to-date design, with a muzzle brake, it could fire a 5.5-kg (12.7-lb) shell to 914.4 m (10 000 yards). But such a performance, allied to a weight of 750 kg (1650 lb), meant that the gun was not very stable at low elevations and tended to jump into the air when fired. It covered an arc from −2° to +70°, but the maximum charge could only be fired above 15°, otherwise the gun leapt into the air and was likely to be damaged. Such a defect annoyed the army and they requested a fresh design in 1940, after having had some experience with the Geb G 36 in Norway.

Below: **A captured 7.5-cm Geb G 36**

I V Hogg

7.5-cm Gebirgsgeschütz 43

Two companies, Rheinmetall and Böhler, were given development contracts for a new gun to be called the 7.5-cm Geb G 43 and they both produced prototypes in 1942. The Böhler design was accepted and four guns were made for troop trials. They fired the same shell as the Geb G 36, to a range of 9510 m (10 400 yards), weighed 580 kg (1280 lb) and stripped into seven loads for pack transport. But in spite of their efficiency (they were much more stable than the earlier gun) by 1943 the army had come to the conclusion that there was no particular need for them and that there were other weapons which were more important. So it was cancelled.

10.5-cm Gebirgshaubitze 40

The Böhler company was an Austrian firm of gunmakers who had already provided the German army with mountain equipment, probably the most powerful mountain weapon ever in use in any country. This was the 10.5-cm Geb H 40, a remarkable weapon. Again it was the result of a competitive contract between Rheinmetall and Böhler.

It was a split-trail weapon with the wheels attached to the trail legs by spring suspension, so that as the trail legs were opened the wheels adopted a peculiar 'toed-in' look. Weighing 1660 kg (3660 lb), it could be towed behind the NSU Kettenrad tracked motorcycle in four trailer loads, towed in one piece behind any convenient truck, or broken into five pack loads for mule transport. It fired a 14.5-kg (32-lb) high-explosive shell to a maximum range of 16 733 m (18 300 yards) and could cover an arc from $-5\frac{1}{2}$° to +71°. It was issued to troops early in 1942 and survived in the hands of various European countries well into the 1960s.

In 1944 Böhler were given the task of developing a 15-cm howitzer, but it is understood that only a prototype was ever built, the war ending before even that could be tested. In the latter part of the war the mountain troops found that recoilless guns gave them all they required in the way of heavy firepower allied to light weight, and as a result there was no further development of mountain guns after early 1943, apart from the Böhler 15-cm design.

COMPARATIVE DATA

Feature	7.5-cm Geb K 15	7.5-cm Geb G 36	7.5-cm Geb G 43	10.5-cm Geb H 40
Weight in action kg/lb	630/1389	750/1654	582/1283	1660/3660
Maximum range m/yd	6625/7245	9150/10 006	9500/10 390	16 735/18 302
Shell weight kg/lb	5.47/12.06	5.75/12.68	5.75/12.68	0.89/1.97
Elevation	50°	70°	70°	71°
Length of gun cm/in	115.5/45.47	145/57.09	163/64.17	344/135.3
Pack loads	7	8	7	5

Infantry Guns

The Rheinmetall 7.5-cm le IG 18, the first new weapon to be issued to the Germany army after the First World War

Infantry guns, as the name implies, are artillery operated by the infantry and not forming part of the divisional artillery organization. Thus, when a company attack runs up against a machine-gun nest, it can deal with it using its own small guns without having to bring the 'professional' artillery into it. There are various arguments for and against the idea: many armies prefer to use mortars, some armies use both, but the principal key to the question lies in manpower. Large conscript armies can usually afford the men to operate infantry guns, while small all-regular armies can not.

German use of infantry guns began during the First World War when surplus mountain guns were given to the infantry. Although mortars were adopted in the 1920s, infantry guns were still demanded, and the first post-war weapon to be issued to the new German army was an infantry gun.

7.5-cm leichtes Infanterie Geschütz 18

Designed by Rheinmetall and issued in 1927, the le IG 18 was remarkable for its construction, being built on the lines of a shotgun. The gun's breechblock formed part of an oblong box in which the barrel was pivoted at the muzzle end. Pulling the breech lever held the breechblock stationary but tipped the barrel up, about its pivot, to expose the breech of the gun for loading. When the round was loaded, the gun breech dropped and positioned the cartridge in front of the breechblock ready to fire. Two variants were also made; the Gebirg IG 18 for mountain troops, having a tubular steel trail instead of the normal girder-section type and thus rather lighter, and the Fallschirmjäger 18F for parachute troops, virtually the mountain model with smaller wheels and capable of being dismantled into four paradrop containers. The standard gun weighed only 400 kg (882 lb) and served throughout the war and saw action on every front.

7.5-cm Infanterie Geschütz 42

In 1940, after the French campaign, the army demanded a more efficient weapon, particularly one which could damage a tank. As a result of this request Krupp produced the 7.5-cm IG 42, a very efficient design using a split tubular trail and fitted with a muzzle brake. Unfortunately there was no production capacity available to make the gun, and the idea had to be dropped. By way of compensation a hollow-charge antitank shell was produced for the IG 18.

7.5-cm Infanterie Geschütz 37

In 1944 the infantry again demanded an improved gun and the Krupp design was brought out once again, but in order to speed up production the gun was allied to some already existing carriages which were in production and could be easily adapted. The first to be taken was that of the 3.7-cm antitank gun PAK 36, and the marriage of this carriage to the new gun became the 7.5-cm IG 37. This weighed 510 kg (1125 lb) and could fire a 5.5-kg (12.2-lb) shell to 5100 m (5600 yards). The gun was fitted with a highly efficient, if somewhat ugly, four-baffle muzzle brake and a semi-automatic breech. The second version, which entered service as the 7.5-cm IG 42, was the same gun mounted on a lightweight split-trail carriage which had been put into production for the 8-cm PAW 8H63 antitank gun. This had a lesser maximum elevation and thus could only reach to 4500 m (5000 yards), though the infantrymen soon realized that by digging holes and dropping the trail ends in, it was then possible to cock the muzzle up a little higher and get the same 5100 m (5600 yards) as the other model. Neither entered service in any great number.

15-cm schwere Infanterie Geschütz 33

The largest-calibre weapon ever classified as an infantry gun by any nation was the 15-cm sIG 33, two of which formed the Heavy Gun Platoon of the infantry regiment. This was a conventional artillery piece weighing 1700 kg (3749 lb) and could be truck or horse-drawn. It fired a 38-kg (84-lb) shell to a range of 4700 m (5140 yards) and was also provided with smoke shells and hollow-charge antitank shells. An unusual projectile issued with this weapon was the 15-cm Steilgranate 42, an over-calibre stick bomb with fins, which was loaded into the muzzle so that the stick entered the bore, the bomb remained outside, and the fins lay alongside the muzzle. Fired with a special blank cartridge this had a maximum range of about 1000 m and was intended as a powerful blast bomb for demolishing strongpoints or wire entanglements. Numbers of 15-cm sIG 33 guns were mounted on self-propelled carriages of various sorts and were extremely useful as assault guns.

I V Hogg

The 15-cm sIG 33 was the largest-calibre weapon ever classified as an infantry gun

COMPARATIVE DATA

Gun	Weight in Action kg/lb	Barrel Length cm/in	Weight of shot kg/lb	Maximum Range m/yd	Elevation
7.5-cm le IG 18	400/882	88.4/34.8	6/13.25	3400/3700	7 5°
75-cm IG 37	510/1125	180/70.77	5.5/12.13	5100/5632	40°
7.5-cm IG 42	590/1300	180/70.77	5.5/12.13	4500/5030	32°
15-cm sIG 33	1700/3750	175/68.82	38/83.80	4700/5140	73°